ENVIRONMENTAL RESEARCH ADVANCES

# RANGELAND ECOLOGY, MANAGEMENT AND CONSERVATION BENEFITS

# ENVIRONMENTAL RESEARCH ADVANCES

Additional books in this series can be found on Nova's website
under the Series tab.

Additional e-books in this series can be found on Nova's website
under the e-book tab.

ENVIRONMENTAL RESEARCH ADVANCES

# RANGELAND ECOLOGY, MANAGEMENT AND CONSERVATION BENEFITS

VICTOR R. SQUIRES
EDITOR

nova publishers
*New York*

**Library of Congress Cataloging-in-Publication Data**

Rangelands: ecology, management, and conservation benefits / editor, Victor R. Squires.
pages cm. -- (Environmental research advances)
Includes index.
ISBN 978-1-63482-504-7 (hardcover)
1. Range ecology. 2. Range management. I. Victor R. Squires.
QH541.5.R3R364 2015
577.4--dc23

2015010478

*Published by Nova Science Publishers, Inc. † New York*

# CONTENTS

# List of Tables

**Chapter 5:**



**Chapter 6:**



**Chapter 8:**

# LIST OF FIGURES

**Chapter 10:**

# List of Boxes

# PREFACE

There are divergent views about what the term "rangelands" might mean. At some stage, there was a dominant view that rangeland referred to "a type of land use". Later, the emphasis shifted to regarding rangelands as "a type of land that has a variety of potential uses", only one of which was livestock husbandry. The notion of *multiple-use* prevailed in many circles and non-conventional uses such as eco-tourism, biodiversity conservation, mining and utilization by military for weapons testing came to be accepted. More recently still, alternative energy facilities such as wind and solar power generation have been installed and installations covering many hectares can be seen in parts of western China, North America, Australia and other regions.

Unlike the notion of a biome that generally lumps together characteristic soil, plant and climate combinations into easily recognized groupings that cover specific geographic regions (e.g., tundra, taiga, savannas or grassland), rangelands are not comprised of distinct ecosystems. Rangelands are spread over every inhabited continent and even sub- Arctic islands like Iceland. They include deserts (both coastal and inland), alpine areas, riparian sites along rivers and lakes. Rangelands may be in the form of vast steppes or extensive grasslands dominated by true grasses (Family Poaceae), savannas with tree overstory or shrub steppes.

Rangelands have been shaped and defined in large measure by the way humans have used them; and human use is always evolving so that during the 20th century in Australia, USA, Argentina the dominant land use was commercial pastoralism (ranching). In Asia and Africa, the dominant land use is semi nomadic pastoralism at the subsistence level. However, this is changing. The initial phase of colonization for commercial pastoralism gave way to a phase of consolidation involving increasing refinement of rangeland management. Technologies that helped increase animal production were developed and implemented. Fencing, for example, gave greater control over animal movements together with the provision of reliable water supplies for livestock. Typically, degradation of natural resources occurred as a result of this intensification. Reorganization of pastoral businesses and institutions occurred later and the cycle began again.

Today, there is a broad spectrum of systems for exploiting rangelands. People have many personal preferences that lower the efficiency of resource use and may even degrade natural resources. This makes it imperative that the definition of rangelands includes both the physical and land use aspects. Large areas in Asia, Africa, Australia and South America are home to indigenous peoples. Despite the development of commercial pastoralism, some indigenous human populations continue to use rangeland resources with their traditional lifestyles and economies.

As economies transition to the market economy and as the pressure of human and livestock populations rise, the emphasis is fast moving toward the maintenance of the basic resources upon which all rangeland uses (and users) depend. There are numerous reminders that factors, other than the biophysical ones, also limit management options and the sustainability of rangeland enterprises, industries and communities. It is now recognized that socio-economic factors are often crucial in determining whether particular technical solutions to rangeland problems can be employed in practice. The long-term maintenance of the resource base is so important. One of the themes in this book is the conservation benefit that rangelands can supply. These are manifested in terms of people's livelihood, biodiversity conservation, land protection, generation and continuity of supply of ecological good and services.

The collection of essays in this book are from many and varied rangelands. Contributions are drawn from the extensive steppes of Central Asia (Mongolia, and western China) from the mountains of western China and Tajikistan to the Mediterranean regions of Spain, and the vast rangelands of central Australia and Argentina. In all of these examples, the three recurring themes of *ecology, management* and *conservation benefit*s emerge to remind us that there is need to realize that our best-intentioned interventions depend on working with land users and other stakeholders. We must learn how to work with people to change thinking and foster the spirit of rangeland stewardship. This becomes increasingly important in the face of global climate change.

The contributors are experienced people with a good grasp of the issues and we hope that the readers (researchers, academic and policy-makers) will benefit from our somewhat eclectic collection of essays.

The extensive bibliography will give access to literature that may be in languages other than English. The book provides insights into issues faced and solutions attempted in less well-known rangeland situations.

***Dr. Victor R. Squires***
International Dryland Management Consultant,
Adelaide, Australia
and Visiting Professor, Gansu Agricultural University,
Lanzhou, China
dryland1812@internode.on.net
squires200@yahoo.com.au

April 2015

# Scope and Purpose

This book is written by experts on the current and latent issues relating to the management of rangelands and brings to the reader a window into this important subject. Technical approaches can address some issues but balancing these is the neglected outcome of social processes. Ultimately it is these socio-economic and socio-political processes related to institutions and governance and engagement with land users that are the hard part, and often a stumbling block, for improving rangeland management and capturing the conservation benefits. Technical interventions such as use of fire and grazing have their role as do rehabilitation efforts to address arrest and reverse land degradation problems and to bring about ecological restoration in lands that cover more than 30% of the world's land surface and provide a home and livelihood to about 1 billion people spread across every inhabited continent. Multiple-use has become a much more important factor in recent decades, especially in the context of global climate change, and the approaches need to deal with issues arising from competing interests requires skills and strategies that are still evolving.

The book is divided into three parts:

## Part 1. Rangeland Ecology: The Underpinnings

This Part of 3 chapters considers important aspects such as the nature and geographical extent of rangelands and the notions of sustainable use from the viewpoint of both ecological principles and the viability of culture and livelihoods of rangeland users. The success of programs to manage rangelands to improve livelihoods, conserve biodiversity and protect the land resource base are considered here with reference to north-west China's vast pastoral areas. The importance of working directly with land users, many of whom are household-based family enterprises who every day make individual decisions that impinge on how the rangeland is managed, to ensure that rangelands are available for use now and in the future is emphasized. The notion of intergenerational equity is explored. Multiple-use has become a much more important factor in recent decades and the approaches need to deal with issues arising from competing interests requires skills and strategies that are still evolving.

## Part 2. Rangeland Management in Practice

There are 4 chapters in this Part all of which focus on experience in managing rangelands. Examples are drawn from widely divergent rangeland situations (Mediterranean Europe, Argentina, Mongolia and Tajikistan). Management objectives are different in each instance but the implementation of management plans and the measures taken to achieve this makes for interesting reading. Whether in the highly-regulated European agricultural scene or in the rural areas of Central Asia (Mongolia or Tajikistan) there are underlying principles to be observed and management interventions that can be 'make or break' such as manipulation of fire and grazing as in the Argentinean example or reforms to land tenure in Central Asia.

## Part 3. Capturing the Conservation Benefits of Better Rangeland Management

There are 4 chapters in this Part. One of the themes in this book is the importance of ensuring that conservation benefits accrue from managing rangelands in a better way. A focus of the Part is on biodiversity conservation and on management of rangelands (and the peoples who depend on them) in alpine regions of north-west China in the context of climate change impacts. Experiences with the aftermath of a culling program on feral camels from Central Australia provide insights into the unintended consequences of past decisions to re-arrange the fauna in Australia.

Finally, an attempt is made to draw together the divergent strands that comprise the management of rangelands to capture the conservation benefits. A number of important themes are implicit in the title of the book 'Ecology, 'Management' and 'Conservation'. Each of these is important in its own right. Clearly an understanding of the ecology of rangelands underpins the management and provides a direction for efforts to conserve the rangelands for the benefit of current and future generations. There is a need to integrate ecological and social sciences and develop a classification of rangelands into appropriate regions/zones in order to address inter-regional questions of production potential, conservation value, resilience and socio-economic viability. Rangelands science has commonly only dealt with biophysical constraints. There have been numerous reminders that factors, other than biophysical ones, also limit management options and the sustainability of rangeland enterprises, industries and communities. It is now recognized that socio-economic factors are often crucial in determining whether particular technical factors can be employed in employed in practice.

# PART 1. RANGELAND ECOLOGY: THE UNDERPINNINGS

This Part contains 3 chapters and considers important aspects such as the nature and geographical extent of rangelands and the notions of sustainable use from the viewpoint of both ecological principles and the viability of culture and livelihoods of rangeland users. The success of programs to manage rangelands to improve livelihoods, conserve biodiversity and protect the land resource base are considered here with reference to north-west China's vast pastoral areas. We emphasize the importance of working directly with land users, many of whom are household-based family enterprises who, every day, make individual decisions that impinge on how the rangeland is managed. It is important to ensure that these decisions mean that rangelands are available for use now and in the future. The notion of intergenerational equity is explored. Multiple-use has become a much more important factor in recent decades and the approaches need to deal with issues arising from competing interests requires skills and strategies that are still evolving.

In: Rangeland Ecology, Management ... ISBN: 978-1-63482-504-7
Editor: Victor R. Squires 

*Chapter 1*

# SUSTAINABLE RANGELAND MANAGEMENT: AN ECOLOGICAL AND ECONOMIC IMPERATIVE

*Victor R. Squires*[*]
Gansu Agricultural University, Lanzhou, China

## ABSTRACT

Sustainable land use is a keystone of sustainable development (SD) of rangelands the world over. SD is a concept that people know about but opinions differ as to what it means in concrete terms. The key question is: "What do we want to maintain, for what purpose and for whom"? Many possibilities exist. These are elaborated in this chapter as is some examination of the relationship between SD and improved land use practices. Sustainability of land use systems should be a high priority in all dryland regions because agriculture (including animal husbandry) is the cornerstone of the economic health and well-being of all them. The meaning of sustainability presents problems, its meaning "in theory" is commonly intuited but "in practice" it is seldom really explained or understood. Likewise, there are the on-going responses to periodic changes in population density, weather patterns, competing land uses, alternative economic uses, and natural condition of resource base for each rangeland site. Perhaps a more appropriate goal is to strive for "more adaptable and sustainable ways of living for the rangeland societies". Maintaining the productive quality of and its resources for continued future use is one of the most important challenges directly confronting the rural population in rangelands but the problems also impact on the wider society (both rural and urban) of all these countries. The intent of this chapter is to identify which players, policies and procedures can best contribute to a more sustainable way of living for the rural societies and ensure that land is used in way that will allow future generations to enjoy the benefit stream that can flow from land that is used in a sustainable way. This gives rise to the concept of *intergenerational equity* and necessarily includes considerations about land stewardship.

This Chapter reviews the government response to the problems of rangeland degradation that is occurring among land users whose incomes hover around the poverty line. The legislative, regulatory and institutional framework has been reformed, and there is now recognition that management of pastoral lands requires cross-sectoral coordination and dialogue with land users. Finally, some ideas are advanced here, including land–

[*] E-mail: dryland1812@internode.on.net

related policy measures, on what can be done to enhance livelihoods, conserve biodiversity and protect land from further degradation.

**Keywords:** poverty, land stewardship, intergenerational equity, heterogeneity, spatio-temporal, nomads, economic rent, local knowledge, ENSO, community, stakeholders

# 1. INTRODUCTION

One-half of the world's countries have portions or all of their land in rangeland environments, many are classified as drylands (Table 1, Fig.1). Climatologists have classified land into several categories based on the ratio of precipitation to evaporation (Table 1) and there are economic implications (see below).

**Table 1. Bioclimatic zones and classification criteria of relevance to dryland**

| Bioclimatic zones | Humidity criteria |
|---|---|
| Hyper arid region | <0.05 |
| Arid zone | 0.05-0.02 |
| Semi-arid zone | 0.02-0.05 |
| Dry Sub-humid Zone | 0.05-.0-65 |

Figure 1. World map showing the principal areas of drylands (too dry, too steep or too rocky for cultivation). Mostly used for pastoral activities (supporting millions of people and their livestock) or as biodiversity reserves, wilderness or wildlife refuges.

But rangeland also encompass uplands, including alpine meadows throughout Asia, Africa, Europe, North and South America and also the desert margins in Africa, Australia

Central Asia and elsewhere. Rangelands are a *type of land* not a *type of land use*. Many rangelands are biodiversity reserves, wilderness areas and forests; others are grazed by vast number of livestock and wildlife. Upland areas are the source of many of the world's most important river systems. For example, the Brahmaputra, Mekong, Yangtze and Yellow rivers arise in the Tibetan region (Squires, Hua and Zhang, this volume).

The drylands and their sub-humid margins represent one-third of the earth's surface and are the home to nearly 40 percent of the world's population (Squires, 2002). There is nothing new in the quest by humankind for permanent ways of using the land entrusted to them. The intent of this chapter is to identify which players, policies and procedures can best contribute to a more sustainable way of living of the rural societies and ensure that land is used in way that will allow future generations to enjoy the benefit stream that can flow from land that is used in a sustainable way. This gives rise to the concepts of intergenerational equity and land stewardship (elaborated below).

Arid and semi-arid rangelands fall at the far end of at least four continua, including those of productivity, rainfall, complexity/heterogeneity, and scientific attention (Stafford Smith 1996). Low productivity tends to lead to large management units encompassing considerable spatial heterogeneity. Globally, rainfall variability increases with decreasing latitude, decreasing annual rainfall, and in regions subject to the ENSO phenomenon, cycles of decade long dry periods. Variable rainfall in arid and semi-arid lands leads to intermittent productivity and makes it difficult to separate the effects of management from the effects of temporal variability (Squires, 2007, Bastin et al., 2012).

Both the rangelands and the people who occupy them are under increasing pressure as pastoral lands lose their traditional role as the major animal husbandry base. This pressure is already being manifested as livestock inventories rise as herders increase herds to compensate for falling returns. In addition, new needs for cash in the market economy and rising demand for red meat from urban dwellers puts additional pressure on agro-pastoral and agricultural regions. Greater grazing pressure is being exerted on smaller and smaller areas of pastureland as the government tries to implement more ecologically-based restoration methods that rely on fencing and grazing bans. The pastoral areas of Africa, Asia and the Middle East present a variety of problems that are both different and similar to those found in other areas of the world where rainfall is more reliable and crops may be grown under rainfed conditions or where stable supplies of irrigation water are available. Resolving them will be difficult and will require policies sensitive to spatial and climatic variability. Because the ecological constraints are different, new cross-disciplinary paradigms that account for interactions between local communities, the natural the resource base, markets and the socio-political environment are necessary (Squires et al., 2008, 2010, Squires 2009, 2012a).

## 2. Will the Future Work in Rangelands?

Rangelands are particularly important as suppliers of ecological goods and services (water, air, etc., Table 2). Any diminution in the supply of ecological services will have negative impacts on "downstream users". In every sense of the word, rangelands are under siege. The siege has already been laid and the battle for the resources they contain has been in place for decades (Squires, 2002b). Most of the ethnic minorities live in the pastoral lands so are geographically, politically, socially on the "edge". So, in addition to being on the edge geographically, there is also a cultural divide. They are on the edge in a political sense too. It is clear that rangeland dwellers have little political clout.

**Table 2. The range of ecological goods and services provided by rangelands is both extensive and essential to people who depend on them for their livelihoods and those downstream users too**

**The benefits people obtain from ecosystems**

| Provisioning | Regulating | Cultural |
|---|---|---|
| Goods produced or provided by ecosystems | Benefits obtained from regulation of ecosystem processes | Non-material benefits obtained from ecosystems |
| •**Food**<br>•**Fresh water**<br>•**Wood fuel**<br>•**Timber**<br>•**Fiber**<br>•**Biochemicals**<br>•**Genetic resources** | •**Climate regulation**<br>•**Disease regulation**<br>•**Flood regulation**<br>•**Water purification** | •**Spiritual**<br>•**Inspirational**<br>•**Aesthetic**<br>•**Educational**<br>•**Recreational** |

| Supporting | | |
|---|---|---|
| Services necessary for production of other services | | |
| ●Soil formation & conservation | ●Nutrient cycling | ●Primary production |
| **Supporting biodiversity** | | |

One of the most fundamental problems confronting people in vast pastoral regions is how to meet the basic needs without destroying the resource base on which they all depend. The history of the rangelands has been one if almost constant deterioration and loss of productivity. The pastoral land degradation reduces rangeland resources, deepens rural poverty and potentially destabilizes ethnic minority areas, and reduces biodiversity. There are several basic drivers of change.

***Population growth***. For example in China there are two sources for this (i) inward migration by peoples from eastern China (mainly Han) and the conversion of vast tracts of the better rangelands into cropland, aided by the damming of inland rivers and creation of artificial oases (ii) natural increase because ethnic minorities were exempt from the one-child policy.

***Climate variability*** within and between years has become less predictable, droughts are more frequent and last longer. Climate change, notably higher temperatures, has caused unseasonal precipitation and given rise to higher intensity events that lead to more run-off and soil and organic matter loss (see also Figure 2).

***Grazing pressure*** (Generally increasing to unsustainable levels) as pastoralists increase livestock inventories as a way to get out of poverty and provide cash for use in the market economy. In many countries livestock are privately owned but the rangeland is in State ownership and land use rights tend to be short term and not very secure (Wang et al., 2010)

This encourages over exploitation (see Squires and Hua, this volume). Both the lands and the people who occupy them are under increasing pressure as pastoral lands lose their traditional role as the major animal husbandry base. This pressure is already being manifested as livestock inventories rise as herders increase herd to compensate for falling returns. In addition, new needs for cash in the market economy and rising demand for red meat from urban dwellers puts additional pressure on agro-pastoral and agricultural regions. Greater grazing pressure is being exerted on smaller and smaller areas of pastureland as the government tries to implement more ecologically-based restoration methods that rely on fencing and grazing bans.

***Livestock density*** (exacerbated in many regions by the imposition of grazing bans and by conversion of rangeland to cropland for fodder/forage production to supplement overgrazed rangelands

***Fire regime*** (extent, intensity, frequency). This is much more of a problem in Africa, Australia (in the dry tropics) and has been shown to a decisive factor (Tothill and Mott, 1985).

It is in drylands (including those where rainfed agriculture is practiced) that environmental degradation is occurring at alarming rates, often leading to desertification (Heshmati and Squires, 2012), and threatening the livelihood of more than 1 billion people. Drylands are diverse in terms of their climate, soils, flora, fauna, land use, and people. No consistent characterization or practical definition of drylands can be made because of this diversity. One binding feature of all dryland environments, however, is their aridity (see Table 1).

The drylands of the world also suffer from the vicious cycle of low productivity (Figure 1 shows relative net primary productivity in the world's drylands), low levels of investment, and, as a result, poverty. Investments, apart from those made for irrigated agriculture activities, are relatively low. Low productivity, low levels of investment, and land degradation often leading to desertification are responsible for regional poverty and income disparities. The poverty and hunger that are prevalent in sub-Saharan Africa is a poignant example of this situation. Other critical problems include the inherent problem of water scarcity, tenure considerations, and ineffective developmental policies (Mortimer, 1998).

Improving this situation requires that a variety of technical and institutional problems be solved. Among these solutions is increasing the level of investments in appropriate agriculture, alternative land use practices, and other appropriate income-generating interventions. Other solutions are designing strategies for risk management and implementing programs for more equitable land distribution and levels of income.[1]

Arid and semi-arid lands exhibit further characteristics that make their dynamics and vulnerability unique, including:

- **modality** (the existence of multiple distinct states),
- **hysteresis** (recovery from disturbance is not always reversible by simply inverting the disturbance event and the path back from a disturbance is often different to the path forward), and
- **divergence** (relatively small changes in initial conditions can result in dramatically different outcomes with time).

[1] ialcworld.org/conference/Pres-pdf/Ffol1px1.pdf.

Environment, at this scale, is also more than a collection of individual phenomena. The emergent ecological properties of landscapes at the scale of human communities are spatial pattern and diversity. Relative heterogeneity of patterns is an important factor in vulnerability to degradation.

Neither the biophysical environment nor human communities are static. Sustainable land use is possible only when the relevant changes and processes within community institutions and environment are spatially and temporally synchronous. Uncoupling in these processes can lead to both social and environmental crises or collapse. Determining relative rates of change is, therefore, particularly important as accelerating environmental changes, including desertification, place new pressures to which the human–environment system must adapt (Vogt et al., 2011).

Despite this complexity, the possible ways in which relative rates of human–environment systems can change are not infinite; there is significant predictive power to be gained from classifying and modeling scenarios at an appropriate level of abstraction. Such a model framework can provide guidance for policy beyond past simplistic recommendations focusing solely on one component of the system or another.

The goal of management in drylands at the community scale (see below) and of policies to support better management, therefore, is the facilitation of adaptive capacity of communities and the understanding of the rates and spatial distribution of environmental change (Squires, and Hua, this volume). Forecasting of environmental change can provide communities with important information concerning the state and trajectory of their environment which may enhance their adaptive capacity.

Sustainability of land use systems should be a high priority in all dryland regions because agriculture (including animal husbandry) is the cornerstone of the economic health and well-being of all of them. The meaning of sustainability presents problems, its meaning 'in theory' is commonly intuited but 'in practice' it is seldom really explained or understood. Likewise, there are the on-going responses to periodic changes in population density, weather patterns, competing land uses, alternative economic uses, and natural condition of resource base for each rangeland site. Perhaps a more appropriate goal is to strive for 'more adaptable and sustainable ways of living for the dryland societies'.

Maintaining the productive quality of and its resources for continued future use is one of the most important challenges directly confronting the rural population in drylands but the problems also impact on the wider society (both rural and urban) of all these countries.

## 3. Not All Rangeland Is Created Equal

Rangelands, especially in the world's drylands vary in their productive capacity, but this is not news. As long ago as 1776 Adam Smith, the Scottish economist, developed the concept of 'economic rent' that was elaborated later by Ricardo 1817 and later still by the German geographer von Thünen (1826). Heathcote (1983) has adapted the economic rent model of von Thünen (Figure 2 [A] and [B]) to show the relationship between land productivity (as reflected in the economic rent that it generates) land use and the influence of rainfall (both the amount and its variability). Market-driven systems depended on access to transport to markets for either live animals or as frozen or chilled meat products e.g., export of meat from

Argentina, Australia or New Zealand to the UK or the Euro-zone. Referring now to Figure 2, we see that in the sequence of zones the grazing of ruminant livestock on rangelands would be the most remote and last zone of productive use. Beyond this, nomadic peoples might still practice subsistence herding. The success of such nomadic systems depended on the unfettered movement to "chase" forage and water and often depended on altitudinal migration to summer pastures in higher altitudes (Kreutzmann, 2012) Jacobs and Schloeder, 2012) or long distance (500-1000 km) migration that may involve crossing international borders (Squires and Sidahmed, 1998 pp.412-415).

Ranching

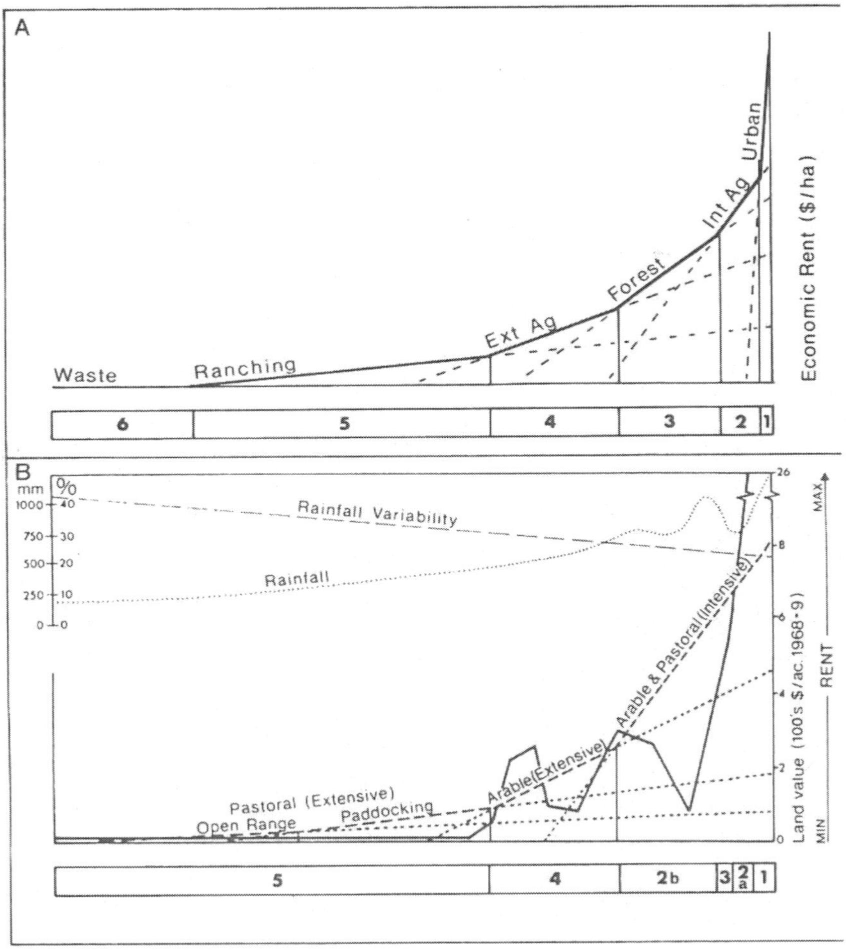


Figure 2. [A] A graph of economic rent as suggested by von Thünen (1826)Hypothetical land use was 1 –urban; 2 – Intensive agriculture; 3- Forest; 4- Extensive agriculture; 5-ranching; 6 – wasteland [B] Land use and land values ranging from intensive land use (far right) to pastoralism on unfenced 'open range'. Rainfall variability is greater as total precipitation decreases (After Heathcote, 1983).

Climatic constraints might reduce the extent of potentially productive areas. As figure 2[B] shows, the pastoral zone is at the arid end of any precipitation spectrum. An annual precipitation regime below 75 mm would probably preclude any pastoral use (nomadic or sedentary) although some camel-based systems might prevail on a subsistence basis.

# 4. Intergenerational Equity

The first priority in all drylands is to maintain and restore the ecological sustainability of the watersheds and rangelands for present and future generations. To achieve and maintain the ecological sustainability of the rangelands while balancing the diverse economic and social needs of rangeland inhabitants along with conserving rangeland biodiversity and watershed values is a major challenge. The wider societies in dryland regions - both rural and urban - have an enormous stake in fostering progress toward profitable, environmentally friendly agricultural and pastoral systems; hence the concerns about the need to foster land stewardship (Squires, 2012a). Regardless of how we define sustainable use of land, it is ultimately the land user who must establish practical, sound practices. Thus, the land users' point of view of what constitutes sustainable land management (SLM) is the most important perception of all. There is a clear need to bridge the gap between production and income objectives of the land users on the one hand, and the long-term objective of preserving natural resources on the other. The challenge that all dryland users face is how to achieve a balance between improved livelihoods, biodiversity conservation and land protection.

There are three intertwined strands of sustainable development that can also have negative impacts on land, often resulting in unwanted feedback loops. Poverty, food insecurity and vulnerability to climatic shocks are likely to remain the major global challenges for sustainable development in the next decades.

The maintenance of land and its resources for continued future use are the most important environmental problems directly confronting the rural population in the world's drylands. But the problems also impact on the wider society of all these countries. Even defining the problem and deciding which outcomes are best is fraught with difficulties, confusion and conflict. Enunciating the principles and practices which can make it work, particularly for rangeland-based industries and related rural populations is the first step. In its broadest sense, SLM is the recognition of the importance of retaining the quality and abundance of land, air, water and biodiversity, and to manage this natural capital in a way that conserves all of its values, be they environmental, economic, social or cultural. Although the definitions are almost limitless, there really are two clear and fundamental elements of SLM – awareness and action. That means recognizing the collective responsibility to retain the quality and abundance of our natural resources. Putting that awareness into action by making the appropriate decisions for how to best use and manage these resources not only for today but for future generations as well. But, implementation of decisions and monitoring outcomes are also required as the land user needs to periodically adapt improvements or changes to emerging conditions in response to past and on-going interventions/actions (Cowie et al., 2011).

The resource base for agriculture, including animal husbandry on arid and semi-arid rangelands, unless managed carefully and replenished continually, will dwindle in its capacity to produce at levels required to meet the demands of burgeoning population and changed market demands from an increasingly urbanized society. Production systems collapse or are forced to change when they become unprofitable to the land user or when they impose on households, neighbors or rural communities (or perhaps whole nations) excessive indirect costs or burdens. These indirect costs arise, for example, from increased frequency and severity of natural disasters such as floods or landslides that are attributable to poor land management.

Rangelands are differentially vulnerable to degradation as a result of their scale as well as spatial and temporal heterogeneity in relation to their attendant land uses (Squires, 2007). The spatial scales of biophysical and social processes are not necessarily linked, although they may partially co-vary. A covariance of properties may exist in that a specific type of community will tend to occupy larger spatial areas and/or seek heterogeneity in some quality of the landscape or land area. For example, nomadic herders tend to require large and, preferably, heterogeneous areas to account for seasonal and other distributional qualities of grass and water. The specific spatial size of the range is associated with the level of vegetation productivity and seasonal differences in it, though not delimited by any single "natural" set of boundaries. Ecological diversity, at least compared to improved pasturage and farmlands dominated by 2–3 species, leads to complex interactions. The limited scientific attention that these lands have historically received, despite their relatively greater complexity, makes locally adapted management systems and knowledge vital for managers (Squires, Hua and Zhang, this volume).

## 5. Local Ecological Knowledge (LEK)

Traditional methods of rainfed cropping and range management practices had been disregarded and undervalued for many decades as regional economic growth was seen as the key with application of modern technology as a means of increasing productivity and way to become modern. 'Modernity' is often associated with abundance, prosperity and a certain emancipation from the hard toil of working the land. Today however, we are witnessing a reversal of this trend and a return towards traditional methods that rely on local technical know-how and practices to restore degraded lands while halting desertification processes that threaten the livelihoods of local people; combating desertification is also a fight against poverty.

There is much valuable information to be learnt from the traditional lifestyles of the people in rural communities the world over, which have evolved over centuries and have thus carved and shaped both the natural and cultural landscape. This traditional knowledge may form the basis of site- and context-specific solutions in combating desertification in the world's dryland areas.

It would be misleading to suggest that all the solutions to the land degradation problems can be found by simply studying and adopting LEK and practices. There are possibilities of rediscovering traditional practice but clear and functional understandings diminish with time. Even when we have the chance to discover these fundamental approaches to sustainability, by witnessing them at first hand, or as a result of patient reconstruction, there still remains the arduous problem of reviving and making acceptable an enlightened approach in a world that is progressively losing its foundational roots in favor of a more dominant and wasteful economy.

It has been argued at several Conferences of the Parties of the UNCCD that 'traditional knowledge' has various advantages and benefits with regard to combating desertification. It is often considered a relatively inexpensive technology (compared with modern technology and computer-driven irrigation schemes, for example, or with drought-resistant genetically modified organisms), and, therefore, represents an affordable corpus of technology for developing countries and their populations. As is well known; many dryland countries that are suffering from land degradation and desertification also rank among the poorest nations of the planet.

Most traditional land-use systems are associated with carefully adapted forms of social organization and regulatory frameworks for access to common resources and ecosystems management, combined with a deep knowledge of the dynamics of the ecosystem over a large territory that is made up of various ecological niches. Such often highly ingenious traditional management systems and cultures have co-evolved over centuries with the landscape and its components, including genetic resources. They are noteworthy for their contribution to biodiversity conservation, sustainable land, water & landscape management, as well as their contribution to ensuring security of food, livelihood, and quality of life. Many provide globally important goods and services well beyond their geographical limits. There is little doubt that LEK systems are a valuable resource for the management of drylands, as they are in other types of ecosystems. However, in many parts of the world, local or traditional management strategies are weakening or losing their relevance, due to rapid changes in their biophysical and socio-economic.

Moreover, LEK has often been highlighted as a particularly 'environmentally-friendly' technology that also blends in with environmental conservation objectives in dryland ecosystems. But, questions have been raised about whether the role of LEK has been oversold.

- Is traditional knowledge indeed an effective means to combat desertification, or do we need to promote modern technologies in the interest of generating higher yields in the fields of agriculture, forestry and livestock husbandry?
- Is it possible to combine both traditional knowledge and modern technologies for effective and sustainable dryland management?

These and other related questions on sustainable dryland management are being considered by various UN agencies, the donor community, NGOs and CSOs.

Pressures are constraining land user innovation and are leading to the adoption of unsustainable practices, over-exploitation of resources and declining productivity, as well as agricultural specialization and the adoption of exotic domesticated species. The result can be biodiversity loss, ecosystem degradation, poverty and loss of people's livelihoods (Squires et al., 2009a). There is a risk of a severe erosion of the diverse base of agricultural and pastoral systems and their associated biodiversity, knowledge systems and cultures that ensure human livelihoods and healthy and resilient environments.

These changes, some of which are driven by processes of globalization, outpace the evolutionary adaptive capacity of the local systems. The focus over recent decades on agricultural productivity, specialization and global markets, and the associated disregard for externalities and adaptive management strategies, has led to a relative and general neglect of research and development support for diversified, ingenious systems. Whilst there is little doubt that LEK systems are a valuable resource for the management of drylands, in many parts of the world, local or traditional management strategies are weakening or losing their relevance, due to rapid changes in their biophysical and socio-economic situation.

Under these circumstances, and viewed from the perspective of the farmers and pastoralist communities, it is not particularly relevant to dwell upon the limits of applicability of LEK versus scientific knowledge. What is more interesting and urgent is how to develop approaches that successfully integrate the comparative strengths of both types of knowledge system. In this context there are three key questions:

- What are the different natures of LEK systems in drylands and of modern or scientific knowledge systems?
- What challenges, constraints and obstacles are there to strengthening traditional sustainable agricultural practices and their knowledge systems?
- What are the principles for an approach to safe-guard traditional management systems for the sustainable use of drylands?

## 6. Dryland Communities: Stakeholders, Actors and Decision Makers

Dryland communities are not homogenous (Seely, 1998). Members of subgroups within a community commonly hold different priorities because of institutionalized differences in access to resources (e.g., livestock ownership, land tenure, user rights and inheritance).

Stakeholders, defined here as community subgroups with divergent interests, may perceive environmental change quite differently, have varying roles in driving that change, and respond to change in differing ways. The process of stakeholder division and negotiation, however, is a community process, one that commonly sits at the center of dryland degradation and remediation. Conflicts arising from differences can drive social change and may be a source of adaptive capacity and resilience, or they may lead to intractable conflict and so limit adaptation.

Communities are comprised of individuals, but individual actions aggregate to impacts and responses beyond the household level (e.g., cropping land similarly in a way that fosters soil erosion). Interactions among individuals and stakeholders generate emergent properties (wherein the whole is greater than the sum of its parts) within a human–environment system that may give rise to unique processes determining and influencing environmental conditions and change. These properties typically involve one or more of the following:

1) institutions that control access to land, water, and other landed resources or the way in which these resources may be used (e.g., usufruct rights or the timing of grazing permitted on common land);
2) means of accessing the State and higher authorities that influence access, production and welfare;
3) mechanisms of access to subsidies, safety nets, and other means of combating the effects of land degradation;
4) local environmental knowledge (LEK): codification of beliefs, knowledge, and ideologies that form the conceptual frameworks within which individual households view any changes in the environment.

The power of the community varies by circumstance. Among peoples only loosely integrated into the nation-state and the market economy, the community is commonly powerful, as it constitutes the traditional level of social organization that integrates individual decisions for the management of resources and resolves conflicts. Recognition of this role neither implies that traditional communities were or are necessarily effective in their management of resources, nor that modern ones are inherently ineffective. Traditional communities have, however, generally had a long period of time over which to adapt to environmental conditions; the relative rate of social and environmental change, therefore, becomes the central issue in determining the appropriateness of various forms of community management.

In arid rangelands, traditional communities are associated with extended kin groups and informal systems of decision making and sanction, like those of nomadic pastoralists, village agro-pastoralists, or rural farmer compounds. Sometimes formal institutions, including village development committees or advisory groups, have been imposed by a national government at the community scale. As these national institutions become influential, community-level influence often wanes. In developed market economies, communities are reflected in the voluntary clustering of land managers, such as pasture user or farmer associations, and the formal institutions of local government. Many communities, especially those transitioning to a market economy, interact with community-level social structure and dynamics, and the influence of the nation-state grows or diminishes (Squires et al., 2009, 2010, Squires, 2012a, Hua and Zhang, 2012, Hua and Squires, this volume, Kurbanova, 2012).

The boundaries of human communities do not usually coincide with those of environmental systems, including catchments, soil types, or vegetation complexes. This can cause conflict, for example, where local government boundaries do not coincide with watersheds, so that incentives for coordinated management of invasive plants and erosion control are reduced (Squires, 2014). Communities do, however, often act to give rise to distinctive biotic landscapes and may create distinctive environmental conditions over their range as, for example, in parts of Western Africa, where communities have produced forest patches in and around savanna villages (Mortimer, 1998).

The community, with its distinctive institutional and ecological position relative to the problem of dryland degradation, is a crucial and largely under-appreciated influence in combating desertification. Community-level processes serve to amplify or attenuate forces of degradation that emerge from differing scales, and community-level interventions have historically proven to both slow and accelerate environmental change. Community-level perceptions provide important alternative views, though ones that may differ among stakeholders. Stakeholders at the community level detect change, explain its causes, project trends, judge consequences, decide how to respond, and implement decisions (Kurbanova, 2012).

## Summary and Conclusion

Sustainability involves the complex interactions of biological, physical and socio-economic factors and requires a comprehensive approach to dryland development in order to improve existing systems and develop new ones that are more sustainable (Squires, 1998).

Several scenarios for sustainable land use have been articulated and most stress the following:

- The inter-connectedness of all parts of a dryland system, including the land users and their families
- The importance of the many biological balances in the system
- The need to maximize desired biological relationships in the system and to minimize use of materials or practices that disrupt these relationships

Recognition of the central role of human activities in processes of implementing SLM has taken a long time to become established (Seely 1998, Reynolds and Stafford Smith, 2002). Supporting research in socio-economic fields, in policy and governance, in economics, in communication and similar fields has gained increasing acceptance but remains far from sufficient (Squires, 2012a).

Rangeland environments must continue to be managed by locally-resident households (Mortimer, 1998, Hua and Squires, this volume, Squires and Hua, this volume). Primary responsibility for making decisions about the management of natural resources in the rangelands will continue to lie where it does now – with dryland households – and for three reasons. First, the intervention capabilities of many governments in the rangeland areas, increasingly debt-strapped, under resourced and dependent on external aid are declining. Second, development objectives for the rural sector have shifted dramatically from overt transformation into new, more ‘efficient’ systems, to providing enabling structures and services, allowing farmers and livestock producers greater autonomy. Third, confidence in the applicability of scientific prescriptions to rangeland management has been badly dented by many project failures, and meanwhile there has been a growing recognition of LEK (though its adaptive capacity, especially when enhanced by education, is probably still under-estimated).

## References

Bastin, G., Scarth, P., Chewings, V., Sparrow, A., Denham, R., Schmidt, M., O’Reagain, P., Shepherd, R. and Abbott, B. 2012. Separating grazing and rainfall effects at regional scale using remote sensing imagery: A dynamic reference-cover method. *Remote Sensing of Environment* 121, 443–457. doi: 10.1016/j.rse.2012.02.021.

Cowie, A L T.D. Penman, L. Gorissen, M.D. Winslow, J. Lehmann, T.D.Tyrell, S. Twomlow, A. Wilkes, R. Lal, J.W. Jones, A. Paulsch, K. Kellner and M. Akhtar-Schuster 2011. Towards sustainable land management in the drylands: Scientific connections in monitoring and assessing dryland degradation, climate change and biodiversity. *Land Degradation & Development* 22 (2): 248-60.

Heshmati, G.A. and Squires, V.R. 2007. A number of ecological propositions and their implications for rangeland management in drylands. pp. 438-442 In: El-Beltagy, A. Mohan C. Saxena and Tao Wang (eds) Human and Nature –Working together for Sustainable Development of Drylands. (Proceedings 8th International Conference on

Development of Drylands, 25-28th February 2006, Beijing, China) ICARDA, Alleppo, Syria.

Heshmati, G.A. and Squires, V.R. 2009. New thinking in range ecology, in Range and Animal Sciences and Resources Management, [Ed. Victor R. Squires], In: Encyclopedia of Life Support Systems (EOLSS), Developed under the Auspices of the UNESCO, *EOLSS Publishers*, Oxford, UK.

Hua, LM and V.R. Squires (In press) Small Holders, the Key to Successful Implementation of Integrated Ecosystem Management. pp. 43-62 In: V.R. Squires (eds) Rangelands: Ecology, Management and Conservation Benefits, *NOVA Press,* N.Y.

Hua, L.M. and D. Zhang, 2012. Engaging with land users: the first steps on long road. pp. 333-356 In: V.R. Squires (ed) Rangeland Stewardship in Central Asia: Balancing Improved Livelihoods, Biodiversity Conservation and Land Protection. *Springer,* Dordrecht

Jacobs, M.J. and C. A. Schloeder, 2012. Extensive Livestock Production: Afghanistan's Kuchi Herders pp. 129-164 In: Victor Squires (ed) Rangeland Stewardship in Central Asia: Balancing Livelihoods, Biodiversity conservation and Land protection. *Springer,* Dordrecht

Kreutzmann, H. (ed) 2012. Pastoral practices in High Asia: agency of 'Development' effected by modernization, resettlement and transformation. *Springer,* Dordrecht

Kurbanova, B. 2012. Constraints and barriers to better land stewardship: Analysis of PRAs in Tajikistan. pp.129-164 In: Victor Squires (ed) Rangeland Stewardship in Central Asia: Balancing Livelihoods, Biodiversity conservation and Land protection. *Springer,* Dordrecht

Li, WJ., and L. Huntsinger, 2011.China's grassland contract policy and its impacts on herder ability to benefit in Inner Mongolia: tragic feedbacks. Ecology and Society 16(2):1.[online] URL: http://www.ecology andsociety.org/vol16/iss2/art1/

Mortimer, M. 1998. Roots in African Dust: Sustaining the rangelands. *Cambridge University Press,* 219 p.

Reynolds J.F and D.M. Stafford Smith (eds) 2002 Global Desertification: Do Humans Cause Deserts? *Dahlem University Press, Dahlem*, Berlin.

Seely, M.K. 1998. Can science and community action connect to combat desertification. *Journal of Arid Environments* 39(2): 267-277.

Squires, V.R. 2002. Desertification: Global Extent In: R. Lal, (ed) Encyclopedia of Soil Science. *Marcel Dekker,* New York

Squires.V.R. 2007. Detecting and monitoring impacts of ecological importance in semiarid rangelands. In: El-Beltagy, A. Mohan C. Saxena and Tao Wang (eds) Human and Nature –Working together for Sustainable Development of Drylands. (Proceedings of the eighth International Conference on Development of Drylands, 25-28th February 2006, Beijing, China). *ICARDA, Alleppo,* Syria pp. 718-23

Squires, V.R. 2009. Processes in Rangeland Degradation, Rehabilitation and Recovery. pp.76-88 In: Victor Squires. Lu Xinshi, Lu Qi, Wang Tao and Yang Youlin (Eds) 2009. Rangeland Degradation and Recovery in China's Pastoral Lands, CABI, Wallingford, UK

Squires, V.R 2009a. Range and Animal Sciences and Resources Management, in Range and Animal Sciences and Resources Management, [Ed. Victor R. Squires], in Encyclopedia of Life Support Systems (EOLSS), Developed under the Auspices of the UNESCO, *Eolss Publishers,* Oxford,UK

Squires, V.R., 2009b. People in rangelands: Their role and influence on rangeland utilization and sustainable management. In: V.R. Squires, (ed) 'Range and Animal Sciences and Resources Management' Encyclopedia of Life Support Systems (EOLSS), Developed under the Auspices of the UNESCO, *EOLSS Publishers, Oxford,* UK, pp. 34-42

Squires, V. (ed) 2012. Rangeland Stewardship in Central Asia: Balancing Livelihoods, Biodiversity conservation and Land protection. *Springer,* Dordrecht 458 p.

Squires, V.R. 2012a Governance and the Role of Institutions in Sustainable Development in the Central Asian Region. pp. 275-303 In: Victor Squires (ed) Rangeland Stewardship in Central Asia: Balancing Livelihoods, Biodiversity conservation and Land protection. *Springer,* Dordrecht.

Squires V.R. 2014. People and Place: Determinants of success or failure to meet Management Objectives in River Basins. pp. 371-396 In: Victor R. Squires Hugh M. Milner and Katherine A. Daniell (Eds) River Basin Management: Understanding People and Place. *CRC Press, Raton Roca*

Squires, V.R. (in press) Unifying Perspectives: Ensuring Conservation Benefits through Better Rangeland Management pp. 181-190 In: V.R. Squires (eds) Rangelands: Ecology, Management and Conservation Benefits, NOVA Press, N.Y.

Squires, V.R. and L.M. Hua (in press) On the Failure to Control Overgrazing and Land Degradation in China's Pastoral Lands: Implications for Policy and for the Research Agenda. pp. 43-61 In: V.R. Squires (ed) Rangelands: Ecology, Management and Conservation Benefits, NOVA Press, N.Y

Squires, V.R. LM Hua and Zhang,D. (in press) Resilience management of Alpine rangeland ecosystems under climate change in North-west China. pp.163-180 In: V.R. Squires (ed) Rangelands: Ecology, Management and Conservation Benefits, NOVA Press, N.Y

Squires, V., L.M, Hua LM., D. Zhang, G. Li 2010. Towards sustainable use of rangelands in north west China. *Springer,* Dordrecht 353 pp.

Squires, V., X. Lu, Q. Lu, T. Wang, and Y. Yang (eds) 2009 Rangeland Degradation and Recovery in China's Pastoral Lands. CABI, Wallingford 264.p.

Squires, V.R. 1998. Sustainable agriculture and pastoralism as a means of combating desertification: which road do we take? *Desertification Control Bulletin No.33* pp 11-15

Squires, V. R. and A.E. Sidahmed (eds) 1998. Drylands: Sustainable use of rangelands into the twenty-first century. *IFAD Series Technical Report* 470 p.

Squires, V.R. and Yang Youlin 2009. How can the next Degradation episode be prevented? pp.247-258 In Victor Squires. Lu Xinshi, Lu Qi, Wang Tao and Yang Youlin (Eds) 2009. Rangeland Degradation and Recovery in China's Pastoral Lands, CABI, *Wallingford*, UK

Stafford Smith, M. 1996. Management of rangelands: Paradigms at their limits. In: J. Hodgson and A. W. Illius (ed.) The Ecology and Management of Grazing Systems. *CAB International*, *Wallingford,* UK

Tothill, J.C. and J.J. Mott 1985. Ecology and Management of the World's savannas. *Australian Academy of Science,* Canberra 354 p.

UNESCO 2005 Combating desertification traditional knowledge and modern technology for the sustainable management of dryland ecosystems: Proceedings of the international workshop Elista, Republic of Kalmykia, Russian Federation 23–27 June 2004 UNESCO–MAB Drylands Series no. 4 UNESCO, Paris.

Vogt, J.V., U. Safriel, G. Von Maltitz, Y. Sokona, R. Zougmore, G. Bastin, and J. Hill 2011 Monitoring and assessment of land degradation and desertification: Towards new conceptual and integrated approaches. *Land Degradation & Development* 22 (2): 150-65.

Waldron, S., C, Brown, and J. Longworth, 2010. "Grassland degradation and livelihoods in China's western pastoral region: A framework for understanding and refining China's recent policy responses", *China Agricultural Economic Review*, 2(3): 298 – 320.

Wang, M.P., CZ Zhao, LM Hua and V.R. Squires 2010, Land Tenure, Problems and Prospects and Reform. pp. 255-282. In: V.R. Squires, L.M. Hua, D. Zhang and G. Li (Eds) Towards Sustainable of Use of Rangelands in North-West China. *Springer,* Dordrecht.

In: Rangeland Ecology, Management ...
Editor: Victor R. Squires

ISBN: 978-1-63482-504-7


*Chapter 2*

# ON THE FAILURE TO CONTROL OVERGRAZING AND LAND DEGRADATION IN CHINA'S PASTORAL LANDS: IMPLICATIONS FOR POLICY AND FOR THE RESEARCH AGENDA

***Victor R. Squires and Hua Limin***
Gansu Agricultural University, Lanzhou, China

## ABSTRACT

People and economic systems in pastoral areas of the five main pastoral provinces in China (Tibet, Xinjiang, Inner Mongolia, Gansu and Qinghai) are complex and people and economic systems are in a process of rapid transition. This chapter provides an overview of the environment and current development trends in the pastoral regions including a glimpse of the people affected most by any conservation or development effort and provide a basic analytical framework for future integrated conservation and development work in the pastoral regions of north and north-west China. The rangelands represent complex ecological and livelihood problems that are exacerbated by the difficulty of devising policy measures that are likely to lead to more sustainable outcomes. The focus is on the reasons why progress to arrest and reverse degradation of the extensive rangelands has been slow and why some initiatives have failed.

## 1. INTRODUCTION

On a global scale, China has around one-eighth of the rangelands (the second largest area of such land in any country other than Australia). Most of these rangelands are situated in the drier and higher regions of north and north-west China. The rangelands are mostly inhabited by peoples of various ethnic minorities (Squires et al., 2010). The combination of remoteness, lack of infrastructure development and low per capita income levels, present complex ecological and livelihood problems that are exacerbated by the difficulty of devising policy measures that are likely to lead to more sustainable outcomes (Brown, Waldron and

Longworth, 2008). The evolving socio-political, ecological, and economic situations in China as a whole are directly relevant to, and to some degree are even determinants of, the success or failure of rangeland management efforts.

Thus, much more than the purely biological must be considered and studied when planning for success (Squires et al., 2009; 2010, Brown, Waldron and Longworth, 2008, Hua et al., 2015). Practical approaches to rangeland management, including attention to biodiversity conservation, imply that proposed solutions must be reasonable, relevant, and useful in the real life contexts of the ecosystem or region where such programs are to be promoted.

No study has yet been published, at least none originating in the environmental or conservation circles that draws together the many interrelated strands, or pieces of the puzzle, that impact rangeland management including biodiversity conservation in China. In particular, no study has yet comprehensively brought together the issues of rangeland ecology, natural resource management, socio-economic development, and biodiversity conservation in this part of the world. Five aspects (economy, ecology, sociology, technology and policy) intersect and interact in the pastoral regions of China. The works of Richard (2005), Harris & Sheehy (2010), Miller & Johnson (2006), Li et al. (2013), and Shang et al. (2014), focus mainly on the nomadic/semi nomadic herders in the Qinghai-Tibet plateau. Other studies have examined the situation in Inner Mongolia (Williams, 2002; and Zhang, Borjigin & Zhang et al., 2007) or Ningxia (Ho,1998) while Squires et al. (2009) have presented case studies on rangeland degradation and recovery attempts for eight of the key pastoral regions in north and north-west China.

People and economic systems in pastoral areas are complex and people are in a process of rapid transition. This chapter provides an overview of the environment and current development trends in the pastoral regions including a glimpse of the people affected most by any conservation or development effort and provide a basic analytical framework for future integrated conservation and development work in the pastoral regions of north and north-west China.

This vast area includes true grasslands, shrub steppe, Alpine rangelands and deserts. Alpine regions are most extensively represented in the Qilian Mountains, on the Qinghai–Tibet plateau and in the Tian Shan and Kunlun Shan of Xinjiang; China's most westerly province (Figure 1). The thematic topics reviewed here include rangelands ecology, pastoralism, sustainability, and biodiversity. The situation in China is analyzed at several different scales in this chapter. We assess the national effort (nation as a whole) and the major pastoral regions in five provinces. The overview of the main development priorities in China's pastoral areas (mostly dependent on animal husbandry) is especially relevant. The specific natural and cultural landscapes receive attention.

The sustainable development of these extremely poverty-stricken, culturally sensitive, strategically important and extremely diverse pastoral areas poses one of the foremost challenges confronting China in the twenty-first century.

The critical roles of local ownership and community participation in conservation efforts (Hua et al., 2015) and the need to meet basic human needs while achieving longer-term environmental needs are discussed. The impacts of several socio-economic policies on rangelands are also analyzed at a regional (provincial) level including a discussion on the role grazing bans and their efficacy. Some of the ecological impacts of fencing and associated changes in livestock grazing patterns on rangeland vegetation are examined.

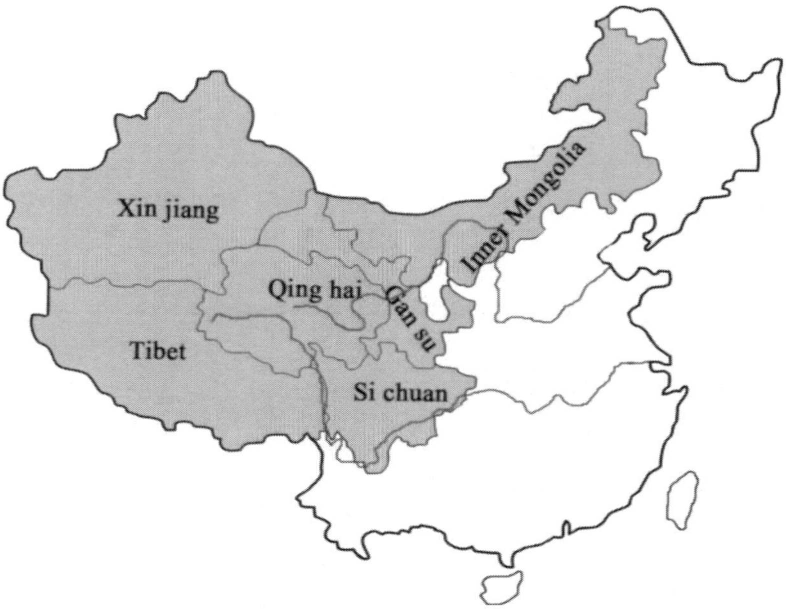


Figure 1. Pastoral areas that are the subject of this chapter are shown in gray.

## 2. The Nature and Extent of Pastoral Lands in Northern and North-Western China

Defining the location of China's rangelands and pastoral regions is not straightforward. The mosaic of pastoral areas, semi-pastoral, and agricultural lands has important implications for how rangelands are used and how policies are developed and implemented. According to official reports, there are 266 counties that are classified as pastoral or semi–pastoral. There are 12 provinces in China that contain pastoral or semi-pastoral counties but only northern and western provinces of Inner Mongolia, Tibet, Gansu, Qinghai,Xinjiang and parts of Sichuan and Ningxia have the majority of the rangelands located in officially designated pastoral counties (Brown, Waldron and Longworth, 2006). These five provinces contain over 75% of the useable rangeland (Figure 2). China has 400 million hectares of rangeland land (42 % of its land area). However, 80 percent is degraded or sanded up[1] due to overgrazing and conversion of grassland to cropland (Figure 3).

The sheer diversity of the rangelands and the overwhelming complexity of ecological and livelihood problems make the task of arresting and reversing land degradation so difficult and help to explain the failure to date, despite enormous effort, to prevent and contain desertification.

[1] Sanded up refers to sand encroachment onto grazing land from unstable dunes or from sand deposition transported by sand storms.

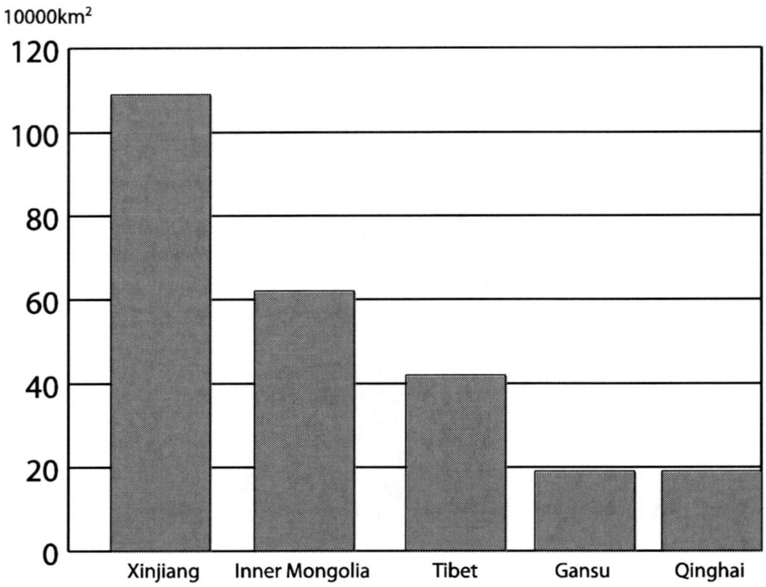


Figure 2. There are 5 main pastoral provincial-level regions in N and NW China that encompass more than 75% of the total grazing lands.

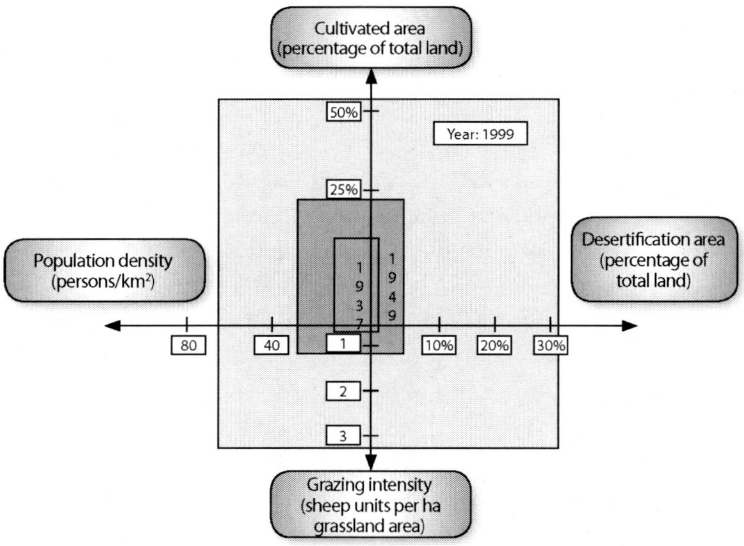


Figure 3. Land degradation has been accelerated by increasing population pressure that led to massive conversion of rangeland to cropland and to increased livestock inventories. In the 60 year period from 1939-1999 in this county in Inner Mongolia over half of the rangeland was converted to cropland (some of it subsequently abandoned) and desertification spread to over 30% of the total area as population density went from about 20 persons per $km^2$ to over 70.

The Chinese government invests 2 billion yuan (250 million U.S. dollars [$US]) a year in fighting desertification, but it is difficult to hit the target that all the country's "curable" desertified land will be restored by 2050.

It would cost at least 238.5 billion yuan (29.8 billion $US) to reach the goal. The annual costs of desertification are estimated at between 54 billion (RMB), equivalent to 7.48 billion $US and 83 billion RMB (11.47 billion $US). So there is financial incentive to try and arrest and reverse degradation.

## 3. Rangeland Degradation and Herder Livelihoods: An Inextricable Nexus?

The focus of the analysis here is how the government of China has dealt with these two problems (degradation and livelihoods) in the pastoral regions. It is recognized that the problems are multifaceted and that different aspects require a tailored response by government. Experience over the past 30 years has shown that a 'National' approach has been applied to allocation of user rights in the belief that 'one size fits all'. Various commentators have pointed to the inappropriateness of the Household Contract Responsibility System (HCRS)[2] to pastoral areas or even in the marginal agro-pastoral interface zone (Williams, 2002).

The HCRS proved successful in agricultural lands in south and east China and was applied nationally without any adjustment to the variable and vulnerable climatic zones that characterize the pastoral zone. The HCRS had led to a highly fragmented and chaotic production and marketing system (Brown, Waldron and Longworth, 2006). The HCRS for pastoralists has been criticized as exacerbating, rather than alleviating, problems of overgrazing in Inner Mongolia (Thwaites et al., 1998, Williams, 1996) and Ningxia (Ho, 2002). In part it is because it led to the creation of a small householder production system (section 6.1). Part of the government's problem is that industry structure designed for other areas of China such as the intensive agricultural regions are not that suitable for the more extensive, geographically dispersed and diverse pastoral system in most of rural China. The HCRS made individual householders (HH) the primary decision makers.

The problem is that individual household level management is not only constrained by institutional factors (that may or may not evolve over time) but also by the realities of land and livestock systems in pastoral areas (that are not likely to change).

Even where long term (up to 50 years) leases have been allocated to individual HH and fencing has partly facilitated exclusive grazing, it is not clear that individual HH management of rangelands and livestock) will lead to efficient and sustainable practices. This is especially so when the principal motive of HH is pursuing self-interest to gain as much as possible from the commons even at the expense of the neighbors and the long term sustainability of the resource base.

Pastoralists have been free to ply the open market since the disbanding of communes and initiation of the HCRS in 1983. After the initial roll out of the HCRS several attempts have been made to tweak the system as a retro-fit. These measures include legislative and regulatory measures (see Brown, Waldron and Longworth, 2008). So far, no measure has

---

[2] HCRS Since the mid-1980s under this policy all livestock and rangeland resources (except the land itself) that originally belonged to the State and were used communally in collectives, were distributed to each householder according to family size, at that time. There was a contract between the government and the householder to produce a quota of produce. Surplus, above the quota, would accrue to the householder. It provided incentive to produce whereas before there was none.

proven to be universally successful, in part due to the attempt to have a universal system that was to be applied across the whole pastoral zone.

Livestock have always played an important part of household activities in pastoral areas. However, the increased number of herders in pastoral areas, especially the agro-pastoral areas, today makes traditional livestock systems increasingly difficult to sustain. Too many livestock producing too little income have placed herders in a cycle of increasing livestock numbers, increasing degradation and declining incomes. Pressure has come from government to 'modernize' in order to provide higher and more sustainable incomes for herders. The primary 'solutions' offered reflect a reductionist approach rather than address the problem within a systems framework. Most government officials see the problem of overstocking being resolved by technological fixes such as improved pastures, breed improvement and fencing.

Herd management measures such as selection and culling for productivity to reduce grazing pressure (Michalk et al., 2010) genuine intra-farm rotational grazing and the production of higher value livestock and livestock products are given lower priority (Brown, Waldon and Longworth, 2006).

The concern that current livestock production systems in many of the pastoral areas of China are not sustainable and that large tracts of China's rangelands are now degraded, were such that the government of China recognized the problem and has enacted various laws and regulations in an attempt to arrest and reverse the problems.

## 4. Legislative and Regulatory Measures

National environmental protection was guided from the 1990s by the *China Agenda 21* Policy framework and by a number of key policy guidelines. The laws provide broad guidelines only and the specific rules and regulations for implementation are established at the lower levels, mainly at the county level.

### 4.1. National Policy Framework

The China Agenda 21 promulgated by the Chinese government has provided a framework for the prevention and treatment of land degradation since the 1990s. Three policy documents provide the state policy basis for the prevention of land degradation.

- The State Council Decision on Environmental Protection (1996)
- The National Ecological Environment Construction Plan drafted by the State Development Planning Commission (1998) – a 50-year plan.
- The Guidelines for the National Ecological Environmental Protection, drafted by the State Environmental Protection Administration – a 50-year plan.

Several policies, from 2000, have supported the Great Western Development Strategy (GWDS). These mainly provided financial support for ecosystem reconstruction projects to prevent land degradation in western China.

These policies do not include sufficient measures to address the socio-economic development needs to the people affected and it is now recognized that further effective regulatory and financial measures are required. A number of policies and laws are relevant to the implementation of rangeland resource management including forestry, environment, land resource management, water and soil conservation, prevention and control of desertification, agriculture, livestock and minority nationalities.

## 4.2. Legislation and Administration

Relevant laws of the People's Republic of China are summarised in Table 1 below.

**Table 1. Legislation and Administration of relevant national laws and regulations**

| |
|---|
| *Land Administration Law*<br>This law controls the allocation of gross land resources to industry and agriculture (agriculture, grassland, forestry and reserves). The General Planning Guideline of National Land Use is the major guideline for land use and purpose regulation.<br><br>*Land Resource Management Law*<br>The law requires that an overall land use plan be prepared for each level of government and with each level complying with the higher level plan. The planning period is 15 years and the current planning period covers 2011-2025. At county level the land use plan preparation is organized by the county government. The County Forestry Bureau (CFB), County Animal Husbandry Bureau (CAHB) and other agencies work closely with the Land Management Bureau (LMB) to produce the plans. Sometimes maps are produced. There are often areas of dispute especially between the CFB and the CAHB, i.e., is it grazing land or is it suitable for afforestation? Land use planning conflicts are resolved at the request of the county government if necessary, by the LMB.<br><br>*Forestry Law (1998)*<br>This law involves various regulations defining forestry resources and forest land. Forest land includes forested land, young forest, nurseries and land suitable for afforestation[3] according to local government plans at or above county level. County level forestry bureaus are responsible to classify the main forest types: shelter (conservation), timber, economic, fuelwood and special purpose forest and submit them to the local government to be approved and announced. A long term forest management plan is prepared at each level. This law requires the compensation to be provided to households that are deprived of forestry resources. |

[3] Forest land in China is loosely defined and shrubland and other rangeland can be classified as forest and come under the jurisdiction of the Forestry Administration.

**Table 1. (Continued)**

*Grassland Law (2002)*
The law defines grassland as grassland, hills and slopes covered with grass, natural grassland and sown pastures. Some counties are in the "agricultural" areas and the grassland is collective, and has not been contracted to individual households (HHs) or group of HHs. Under local legislation some grassland law is applied to brush (grassland and thin forest) so there is potential for conflict between Bureaus at local levels.

*Agricultural Law*
This law regulates agricultural activities and provides the legal basis for contracting farmland to households, farmland protection and farmland infrastructure development. Agricultural land includes degraded and abandoned cropland.

*Water Law (2002)*
This law provides the legal criteria for the regulation of the planning, utilization and protection of surface, underground and aerial water resources.

The main laws against land degradation in China are:

- Water and Soil Conservation Law (1991, 2011)
- Prevention and Control of Desertification Law (2001)

All of the laws listed above should coordinate with the two laws mentioned below. The Ministry of Water Resources is responsible for enforcing the Water and Soil Conservation Law and the State Forest Administration (SFA) is responsible for enforcing the Prevention and Control of Desertification Law.

Special attention needs to be paid to the *Water and Soil Conservation Law* (1991, but revised with effect from March, 2011) and to the *Prevention of Desertification Law (2001)* because these two pieces of legislation are the main ones specifically about land degradation and its prevention and control.

*Water and Soil Conservation Law 1991* (as amended in 2010) is formulated for the purpose of the prevention and control of soil erosion, the protection and rational utilization of water and soil resources, the mitigation of disasters of flood, drought and sandstorm, the improvement of ecological environment and the development of production.

*Prevention of Desertification Law (2001)* This Law was enacted to prevent land desertification, rehabilitate desertified land, maintain eco-safety, and promote sustainable economic and social development.

But it is the *Grassland Law*[4] that is most relevant to the discussion in this chapter. The *Grassland Law* began to be implemented after June 1985 and was later amended in September 2002 to take affect from March 2003. The roll out of this legislation was variable

[4] Grassland in China is catch-all term that corresponds to the term rangelands that is more commonly used in international literature. It includes true grassland (prairie) steppe, meadow and even shrubland.

between provinces and in some was delayed by 5-7 years despite the fact that the national and provincial level policies, laws and directives in the *Grassland Law* must be followed and obeyed. Guidelines on the rules and regulations might be given from each provincial level Bureau as appropriate the local circumstances. Technical guidance and training is provided by the Ministry of Agriculture's Animal husbandry Bureau (AHB). These provide direction for the county policies that guide the day to day implementation and enforcement, especially of grazing bans.

## 5. Provincial Grassland Policy, Law and Grazing Ban Implementation

Provincial laws and directives must be obeyed and should be based on the national laws and directives. Provincial laws, rules and regulations are constituted according to the local conditions of each province. Implementation of the grassland law is the responsibility of the Animal Husbandry Bureau through the Grassland Division and its Grassland Monitoring and Supervision Station.

A major national directive on grazing ban protection (released in most pastoral provinces in 2001) specified 10 land types/land uses that should be protected by grazing bans. The specifications of these ten areas are drawn from several national laws and have implications for overall land use planning. Provincial laws and directives mirror the national level. Detailed rules and regulations are set mainly at the county level. In areas classified as *agricultural* long-term user rights (URs) are uncertain, especially on legal cropland for which the farmers will have two UR certificates, one for crop and one for so-called 'forest land'. Implementation of the grazing ban is highly variable ranging from good to poor due to a range of factors including the lack of comprehensive rules and regulations, the lack of compensation for the loss of traditional and/or legal grazing rights, boundary disputes between villages and title disputes.

There are often no county-level rules and regulations covering the general parts of the grazing ban protection directive, i.e., management of catchments and eroded land, water reserve areas, degraded grassland and "other areas that need to be protected". These would cover areas such as degraded bare forest land and overgrazed grassland. At township and village level there are "rules" for local behaviour including logging, mining and illegal grazing including penalties but illegal grazing refers mainly to forested areas. There are few detailed provincial or county land ownership and classification and land use plans and maps on which to apply a clear set of grazing ban rules and regulations. The Land Management Bureau is responsible for preparing 15-year land use plans. The current plan runs from 2011-2025 a new phase of mapping is under way. Complex land use plans and grazing bans require multi-agency implementation and coordination. There are inadequate resources to police, even in the clearly demarcated grazing ban areas.

Responsibilities related to grassland and implemented by the AHB are:

- Implement and monitor the *Grassland Law*
- Supervise and monitor the grassland in the province

- Supervise and monitor for grassland management, protection, establishment and utilization
- Issue and register grassland titles and help the government at or above county level to resolve disputes on grassland title
- Determine the grassland carrying capacity
- Prevent grassland fires
- Execute penalties according to the *Grassland Law*.

## 5.1. Grassland Grazing Ban Planning Process

Following the national directive on grazing ban protection in 2001 and the amended *Grassland Law* in May 2002 further instructions were given to the provinces in 2003 from the national government and the implementation of grassland grazing bans commenced in 2004. Based on the national grassland law and the national directive on grazing ban protection there are several guiding documents in each Province. Commonly there is planning for the protection, establishment and utilization of grassland but in most Provinces there is no clear implementation planning, targets and financing for grazing bans, especially in the transitional agro-pastoral areas. All counties were expected to follow the Provincial directive on grazing ban protection from 2001. The several land type/land use areas of the directive that might include grassland, e.g., management of catchments, water reserve area, degraded grassland and "other areas that should be protected".

According to information from the national grazing monitoring centre, over 71 million ha of pastureland will be covered by the temporary total grazing ban for several years. This means that conversion to indoor raising of more than 20 million farm animals must occur in the period 2005-2010. This further exacerbates the need to grow fodder on land that was used for food crops and leads also to encroachment onto pastureland that is plowed for crop production. So a vicious cycle is set in train (see below). China spent around 26 billion RMB (3.6 billion USD) on restoring 660 million hectares of pastureland. There is little doubt that grazing bans work but are they cost effective and are they fairly applied? In pilot projects, after three years of the grazing ban in Inner Mongolia, the vegetation cover increased from 20 percent to over 60 percent. Local sandstorms have declined appreciably, and forage yield has risen from 450 kg/ha to 1,500 kg/ha by virtue of the grazing ban. The ban is the right policy, but the sustainability of its positive results will be in jeopardy if the ban is modified without introducing a sustainable pasture management system, and unless livestock farmers are supported in generating a stable income.

Grazing ban implementation is the responsibility of the Grassland Monitoring and Supervision Station (GMSS) of the AHB. The lower levels of government may adjust the rules and regulations to suit the local conditions. The Grassland Monitoring and Supervision Station of the AHB has the responsibility to monitor the condition of the grasslands and supervise any grazing bans.

Responsibilities related to grassland and implemented by the County-level AHB mirror the provincial responsibilities. Supervision is the main responsibility in counties. Very few issue grassland user rights (titles) or try to solve disputes. The counties do not have the funds to survey carrying capacity.

The GMSS of the AHB in each county is also responsible for implementing the grazing bans related to grassland in the agricultural zones, although some land is classified as 'forested' and falls under the control of the Forestry Bureau.

It is noted that in the semi-agricultural and pastoral areas it is difficult to implement the grazing bans because grazing is a traditional way of life. There are limited funds for supervision and there are frequent conflicts between the government supervisors and the herders (Dong et al., 2007).

The existing national laws and grazing ban protection directive provide an adequate framework in which to formulate the grazing ban rules and regulations but implementation requires multi-agency planning, implementation and monitoring, preferably under a lead organisation or special institution. Seasonal or total grazing bans imposed abruptly by the State have actually created the uncertainties that HH have about their use of rangeland assigned them under the HCRS. Under these circumstances HH may well seek to make the most use (grossly overstock) these rangelands before their use is further modified. Thus the institutional uncertainties may have increased the incentive to overgraze in recent years.

There are few comprehensive official land use maps and plans on which to base grazing ban rules and regulations. There are only inadequate and incomplete rules and regulations at county level and below. Little guidance has been provided from the provincial level. There is no short or long-term compensation for farmers affected by the grazing bans in 'agricultural' counties. There are no provincial and county funds for rehabilitation afforestation in most places unless the land is under the national program 'Grain for Green' (Wang et al., 2007) where farmers are compensated for planting trees or grasses on marginal cropland. Box 1 sets out some ideas about 'where do we go from here' if we want to get serious progress in making grazing bans part of the solution and not part of the problem.

**Box 1. Key components of a successful grazing ban program include:**

Comprehensive land use planning and mapping for the next land use planning period 2011-2025

Removal of any legal conflict between forest land and grassland, and reconciliation of boundaries between counties, townships and villages

Comprehensive clear, transparent and uniform multi-disciplinary rules and regulations regarding grazing bans and other activities

Adequate compensation for grazing bans including community infrastructure, farming systems and vocational training, and long-term custodial payments for protecting the environmental resources on behalf of downstream beneficiaries

Participatory planning of an overall village and sub-catchment environmental and socio-economic plan

Adequate resources for the various agencies for planning, implementation, enforcement and monitoring

Take steps to remove uncertainties created by sudden and unexpected imposition of grazing bans

Despite widespread imposition of year-long grazing bans in many areas and seasonal grazing bans in others and the expenditure of millions of dollars on fencing but there is little evidence of long-term rangeland improvement (see below).

A failing of the current grazing policy is that it does not allow for adequate rangeland recovery or provide enough flexibility for the herders to move livestock. Allowing herders flexibility and mobility is a very useful tool for management of livestock in these highly non-equilibrial environments.

Government programs have attempted to provide pastoralists with seasonal grazing pastures, but have also required herders to use specific pastures at specific times.The national program "Return Grazing land to Grassland" (RGLGL) has involved massive fencing programs some of which have worsened the situation by accelerating degradation (Williams, D., 2002, Williams A., 2010, Squires et al., 2009). Commonly enough, herders and local land users were not consulted about the location of fences *vis a vis* water point access or migration routes. In all too many places fences were cut and trespass grazing was rife. Even where government subsidies were accepted as part of the trade-offs sought under the RGLGL herd numbers have continued to rise. Part of this rise has come from a response to increasing demand for red meat, especially beef, and herd size has increased to cater for it. Not all of the animals are on the rangelands and feed lotting on small scale has occurred. Dalantai (2008) reported that the proportion of livestock on the rangeland in Inner Mongolia has decreased by 30%. A much higher proportion is now penfed.

In a study by Hua et al., 2015, a five–year comparison of three governance scenarios in Gansu it was found that there was little difference in rangeland health between the area under the RGLGL and the unregulated use grazing of common land (the Control treatment). There were considerable benefits in terms of both rangeland health and household income under a system involving cooperating (Allied) HH combining their herds and grazing a similar sized area under a herder-devised management regime where entry and exit dates were strictly adhered to, herd numbers were regulated and trespass grazing was eliminated.

Only under such systems where there is strict adherence to a previously agreed management regime can individual herders feel confident enough to reduce their herd size and practice more rigorous culling and selection. In the paper by Hua et al., 2015 it was shown that reduced herd size contributed to higher returns to HH. The principle of *more with less* (Michalk et al., 2010) was amply demonstrated.

As referred to earlier, the principal motivation of herders is a selfish concern about getting advantage over their neighbour and they see the key to getting rich is to have a larger livestock inventory. Only when this mindset changes can progress be made. All the regulation, all the subsidies will count for nought unless there is a fundamental change in thinking. Evidence that this can occur is in the success of Pasture User Groups (PUGs) throughout the pastoral zone where HH devised rangeland management and development plans and agree to abide by the locally-developed rules about restricting herd size, honoring agreements about rotational grazing and so on. On the Hulunbeir grasslands in Inner Mongolia, where fencing of individual land areas was done decades ago there is a movement to now remove many of the fences to allow cooperating (or allied) HH to control use of the combined area over which they hold User rights and to manage the area as a whole (Zhang, 2006, Zhang, Borjigin and Zhang, 2007). This allows more effective rest rotation and spreads the grazing pressure. This arrangement, although involving collective action, is not like the old collectives where all the land and the livestock were government owned and controlled.

The new Allied Household system gives individual herders some dignity and allows them to make informed decisions about how many livestock and which type to keep. They can use collective bargaining to get better prices for costly inputs (veterinary supplies, petroleum fuel etc.) and, when selling, a better price for wool, cashmere or livestock for slaughter.

## 6. How Can We Prevent or Reverse Desertification and Land Degradation (DLDD)?

Factors contributing to the degree of difficulty faced and reasons for a failure to prevent overstocking and accelerated rangeland degradation are set out below.

1. The desert and desertified lands are so large, taking up 27.4% of the nation's land area, therefore the task of arresting and reversing land degradation is arduous and not always successful.
2. With China's huge population and relatively slow economic growth (in the worst affected areas) there has been a reckless and over-zealous pursuit of economic benefit e.g., conversion of rangeland into cropland, overgrazing and excessive excavation of minerals and building and road-making materials etc
3. Input for combating desertification, especially in rangelands, is seriously insufficient and it obviously does not match the heavy schedule of tasks that need to be carried out.

Apart from those constraints a much more important barrier relates to the almost universal small holder production system that came about after decollectivization and the implementation of the HCRS.

### 6.1. The Small Holder Production System Model – a Barrier to Progress

Household-based business is the main production model in China's pastoral industry. In this model, use of rangelands is extensive. It is unlike the intensive farming practices of most of China's agricultural zone in the better-watered east and south of the country. Such a 'small but all-round' production model is not helpful in increasing economic gains. To exacerbate the problem the HH property is split into shares when the children grow up or the sons set up their own family business on their share of the original land allocation. Because all suitable land was allocated after decollectivization there is no land left. It is true that under the HCRS there was an opportunity for a HH to benefit from their labors and this led to higher productivity but the implications of expanding populations of land users' offspring had not been thought through. Under the HCRS arrangements the original land allocation remains protected but subdivision of what might have been enough land to support a HH becomes woefully inadequate as generational changes occur. For example in Sunan county in Gansu, the number of HH increased from 1508 in 1954 to 7818 in 2010 which means that the rangeland area for each HH has become smaller while total rangeland area remains the same. With such a small rangeland area available to them it is difficult for individual HH to improve management such as zoning for rotational grazing, spring deferment or rangeland improvement such as re-seeding. Such a business model has resulted in low HH income that impacts on livelihoods and health and well-being and leads to unsustainable development. A survey of HH in Sunan County, Gansu (Hua and Michalk, 2010) demonstrated that each HH needs to maintain flock/herd of about 200 sheep units in order to meet the inescapable costs that must be paid.

To make matters worse the strict 'one child policy' does not apply to ethnic minorities most of whom live in the pastoral zone within the 5 provinces that are the subject of this chapter. The government response to this dilemma of burgeoning population and limited land is to encourage people to migrate and seek work as migrant labor. For example, statistics from the second agricultural mass survey shows that a rural population of 2.478 million, which represents 20% of the total workforce of Gansu, has migrated from rural areas (Hua, 2009). Statistics from other regions suggest a similar story. Even those who stay, supplement their income with seasonal labor or re-train as drivers, mechanics or trades persons. Many have taken advantage of the government sponsored "Sunshine program"[5] that provides training for rural labor to become qualified to take semi-skilled jobs. A feature of the movement of people, especially young people from the pastoral areas, is that it is often the more educated who have the capacity and motivation to move. This has potentially serious consequences at a time when innovative and motivated people are required on the land to manage it better.

The out-migration of HH and their access to non-pastoral and off-farm employment opportunities in these remote pastoral areas is far less common than is the case in Han Chinese agricultural areas of China because of deeply embedded culture, language barrier[6], education and skill level and other constraints. Thus development options in pastoral areas and livelihood options for remote pastoral HH remain inextricably linked to the rangeland and livestock activities. Government officials at all levels assess development options based on this premise.

As a result, increasing emphasis needs to be devoted to designing better rangeland and livestock systems and hence development strategies (Kirychuk and Fritz, 2010, Michalk et al., 2010).

## 7. Preventing Rangeland Degradation Is Better (and Cheaper) than Rehabilitation

Effective prevention of desertification requires both local management and macro-policy approaches that promote sustainability of ecosystem services. It is advisable to focus on prevention, because attempts to rehabilitate degraded areas are costly and tend to deliver limited results. Measures designed to reverse desertification and land degradation (DLDD) are outlined for a variety of situations from the deserts of China, Africa and the Middle East to the high rainfall regions in Asia (Heshmati and Squires, 2013).

---

[5] Set up in 2004 by Ministries of Agriculture, Finance, Labor & Social Security, Education, Science and Technology and Construction.

[6] There 56 different ethnic minorities in China with a plethora of languages.

There are two important questions: why are actions needed and what actions can be taken to prevent desertification and land degradation?

### 7.1. Why are Actions Needed?

Desertification and land degradation (DLDD) lead to loss of productivity impacting livelihoods and economic viability leading, in the worst situations, to deepening poverty. Ecological stability is another consequence that affects sustainability in the long term. At the end of 2009, China had a total desertified land area of 2,623,700 $km^2$ making up 27.3% of the national land area. Desertification mainly occurs in the 5 provinces that are the focus of this chapter (see Figure 1 and Table 2).

### 7.2. Assessing the Benefits of Rangeland Rehabilitation

Land degradation takes place and has its impacts over a variable time frame, most often in non-linear and dynamic ways. Degradation reversal can have positive effects on rangeland ecosystems. Therefore, we should compare the changes to the ecosystem services 'with' and 'without' interventions.

We know that reduction or disruption of the flow of ecosystem services in degrading rangelands could impact rural HHs income and exacerbate loss of land productivity as land users try to intensify production or move it to more marginal areas. Core ecosystem processes and services change over time but the change is often accelerated in the degradation and recovery phases (Squires et al., 2009). Therefore we should consider the key ecosystem services changes as including both flow and stock changes and that these are likely to be different where interventions are instigated to restore ecosystem function (Figure 4).

Many resource economists are now applying tools such as Cost Benefit Analysis (CBA) or Cost Effectiveness analysis (CEB) to evaluate the economics of rangeland ecosystem degradation and recovery (UNCCD, 2013). One approach is to use data on HH income and the level of expenditure associated with rangeland improvement interventions.

**Table 2. The distribution of desertified land by province in each of the five that are most seriously affected. These lands comprise 95.5% of the total desertified land in China and 93.7% of all sandified land**

| Attribute | Province | | | | |
|---|---|---|---|---|---|
| | Xinjiang | Gansu | Tibet | Qinghai | Inner Mongolia |
| Area of desertified land | 1,071,200 | 192,100 | 432,700 | 191,400 | 617,700 |
| Area of sandification | 746,700 | 119,200 | 216,200 | 125,000 | 414,700 |

Sandification and Sandified are terms that refer, respectively, to a process whereby sand encroaches from unfixed sand dunes upwind of a site leading to sandified land where vegetation is buried beneath sand.

Multiple regression assumes the relationship between rural HH income and land productivity (using rural HH income as a proxy) and the rural HH production inputs (such as land, labor and capital etc.), the specific rangeland degradation control measures or other technology transfers, and the characteristics of HH and village condition has the form presented in Eq. 1, 2 below. We could use the following estimation to predict the changes in rural HH income or land productivity.

$$R\mu = e^{ae}\Pi^{j}_{j=1} X^{aj}_{u}\Pi^{K}_{k=1} Y^{\beta k}_{k}\,\Pi^{M}_{m=1} Z^{\gamma m}_{mt}\,\varphi u \qquad \text{(Eq.1)}$$

$$\text{Or } InR_u = \alpha_{0+}\Sigma^{j}_{j=1} a_j\, InX_{u\,+}\, \Sigma^{K}_{k=1}\beta_k\, InY_{k\,+}\Sigma^{M}_{m=1} Y_m\, InZ_{m\,(}\,\Sigma^{M}_{m=1} Y'_m Z'_{m\,+}\,\varepsilon_u \qquad \text{(Eq.2)}$$

where: R is a sample household income or land productivity: $X_{1...}X_k$ production inputs: $Z_{1...}Z_m$ are the characteristics of HH, natural, and village conditions; $Y_1...Y_k$ are the rural household's engagements in the specific degradation control measures[7]; *i* is the *i*th HH; and *t* is the year (*t*=1,2,n). The stock changes could be calculated by using growth models for different natural resources e.g., trees, grasses, livestock.

We could compare the flow changes and natural resource stock changes in ecosystems that had interventions (or none) as the following for time *t*. The benefit change of the *i*th rangeland ecosystem service for time *t* is $s_{it.}$

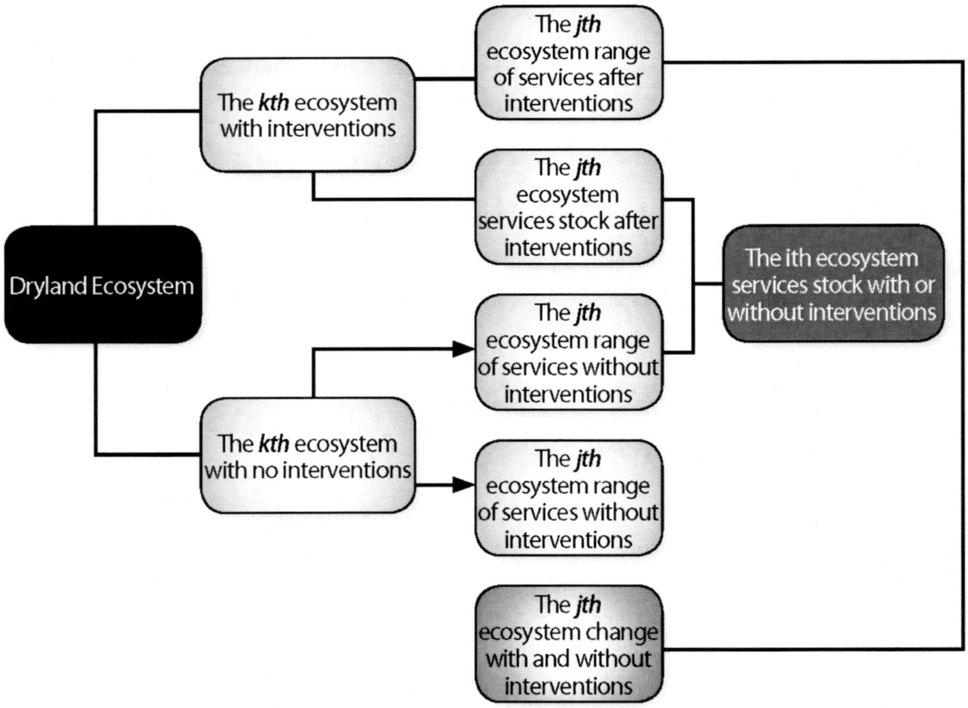


Figure 4. Cost benefit analysis: Conceptual framework for estimation of dryland degradation control.

[7] For example, controlled grazing, spring deferment, planting fodder, building warm pens for winter housing of livestock, fodder conservation, etc.

Such analysis is part of the broader push to justify eco-compensation to land users whose actions improve the flow of ecological goods and services and payments for ecological services (PES) by beneficiaries. There is growing support for applying eco-compensation payments to HH in pastoral areas, especially in the uplands, as a way to incentivize land users to take a stewardship role and benefit not only themselves but other offsite users who depend on the flow of goods and services (Squires, 2012). If widely adopted throughout China's pastoral lands it could do much to improve the situation for HH livelihoods but also reduce the frequency and severity of dust and sand storms, floods, landslides, improve water quality, conserve biodiversity and foster eco-tourism.

## 7.3. What Actions Can Be Taken to Prevent Desertification and Land Degradation?

Because DLDD is an on-going process some actions may be more urgent than others and temporary or stop gap measures may need to be employed. However, since DLDD processes include many issues – bio-physical, socio-economic and cultural and spiritual aspects (Zhang, Borjigin and Zhang et al. 2007) they necessarily need political, legislative, institutional, and policy reforms. Once these are in place, administrative follow up and enforcement of regulations is required.

In short, creation of an enabling environment is a first step (Squires 2012).The principal benefits to society from arresting and reversing land degradation will be higher household incomes, increased productivity of farming systems, conservation of biodiversity, increased carbon sequestration, and fewer floods and landslides.

More sustainable land use practices can help but sustainable farming systems and practices will be adopted when, and only if, they offer farmers a convincing opportunity to earn higher profits than from any other systems. There really are two clear and fundamental elements of "better land *stewardship*" – *awareness and action*. That means (i) recognizing our collective responsibility to retain the quality and abundance of our natural resources and (ii) putting that awareness into action by making the appropriate decisions for how to best use and manage these resources not only for today but for future generations as well. This has aspects of improved stewardship but needs to be seen against a background necessity for livestock keepers and farmers to provide a livelihood. Finally, we must remember that current systems often are used because they were used the year before, are proven, reduce short-term risk, and require modest investment and a low level of management skill and equipment.

### *7.3.1. Land Tenure Reform - An Economic and Social Imperative*

Reform of policies for rangeland user rights and the prevailing land tenure situation, especially in the interface zone between farming and pastoralism has been identified as a major cause of accelerated land degradation (Williams, Wang and Zhang, 2009). A key factor in determining the level of rangeland degradation is property rights, or strictly speaking user rights, because land is owned by the State. User rights govern stocking rate limits, fees and other obligations.

They affect the incentives of rangeland users to pursue particular livelihood strategies and manage rangeland resources. Reforms need to be put in place to encourage livestock owners to stock at sustainable levels and invest in rangeland at a suitable level to ensure their use into the future. Moving toward socially optimal resource use requires property rights that are complete, enforceable, transferable and that account for non-market values (Halimova, 2012).

#### *7.3.2. Poverty Alleviation and Land Degradation Control in N and NW China*

The environmental consequences of rangeland degradation diminish the production base for both agricultural and industrial development and increases the potential for poverty which in turn fosters a cycle of resource exploitation. Combining the degradation control programs with poverty alleviation is a fundamental principle that has been slow to be realized but which of late has become the government's approach. The State Council *Proposals on further promoting development of the western region* and the *Development Plan of Poverty-alleviation program for Rural areas 2001-2010* both included policy elements to eradicate poverty. Combining a poverty-alleviation program with a degradation control strategy will have positive results for human and environmental management.

## Conclusion

China, with largest desert and desertified land area in the world, faces tremendous challenges in the process of combating desertification due to the combination of factors that contribute to the problems and which impede their rectification. As a rapidly developing country there has been a rush to exploit resources that will fuel economic growth and in some cases this zeal has been at great cost to the environment. Since the 1990s there has been a better appreciation of the magnitude of the problems both biophysical and socio-economic. There is need to tackle land degradation through (i) an integrated, participatory, and cross-sectoral approach aimed at the root causes, and resolve inherently conflicting policies (ii) evolve effective mechanisms to coordinate policies, programs and actions by various sectoral agencies operating in areas of agricultural and rural development, land, forestry and water management, environmental protection, finance and planning.

Earlier, the socio-economic dimensions were almost completely ignored when 'ecological construction' was the key phrase that epitomized the notion that "man was dominant over nature".

As a large developing country, China faces tough challenges ahead and has a long way to go, even beyond the 2050, the deadline set by the Central government for repairing the ecological damage done, mostly since then formation of New China in 1949. Realization has come that there is need for a new approach. The more recent Chinese approach is characterized by: the implementation of radical measures like:

- the wide-scale grazing ban;
- the integration of desertification control into regional development in the west of the country;
- wind erosion control in all climatic zones affected;
- improved water resources management and integrated management of smaller watersheds;
- a stronger emphasis on prevention.

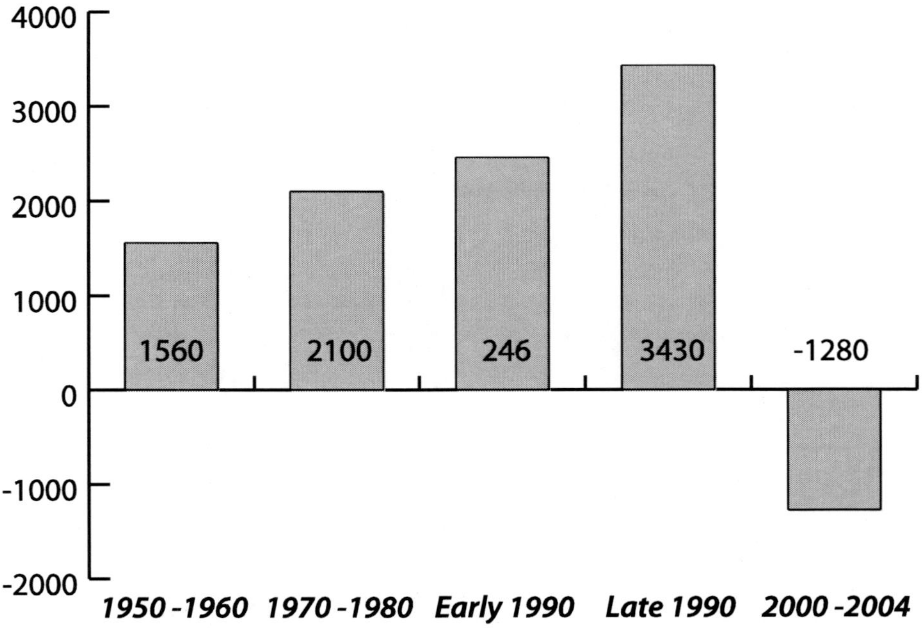


Figure 5. Trend and dynamic of sandy land desertification in China is the 50 years to 2005.

With this approach, for the first time since the founding of the People's Republic of China, the constant spread of desertification has been halted, especially in sand fixation, oasis and infra-structure protection. Much of the success has been in the sandy lands (Figure 5) where in comparison to the situation in 1999 there has been a reduction of 6416 $Km^2$ by 2004. Further gains have been made.

However, activities to regenerate vegetation by excluding people and animals coupled with conversion to different management practices run up against limitations wherever they infringe on cultural values and traditional ways of life (e.g., those of the semi-nomads in Tibet, Xinjiang and Inner Mongolia), and where the financing of protection and management conversion measures is not guaranteed. It will thus be critical for long-term success to motivate the affected land users and herd-owners as well as private investors by offering suitable incentives to play a far greater part in the regeneration of vegetation cover and conversion to environmentally benign and sustainable land use.

Very soon the efforts to achieve sustainable development in the regions threatened by desertification could be threatened still further by climate change. The race cannot be won by battling against nature. The harmonization of society and nature is also an objective of the 12$^{th}$ 5-Year Plan.

In desertification prevention and control, not all of which involves grazed rangelands lands, China will undoubtedly continue to achieve major success in some areas. Yet it is a task of unprecedented proportions and there is little expectation that the task will be completed any time soon for the reasons outlined above. There is recognition now too that there are few management options available to land users, especially herders. Those that do exist fall into two categories (Figure 6).

The government proposes four strategies to deal with the problem.

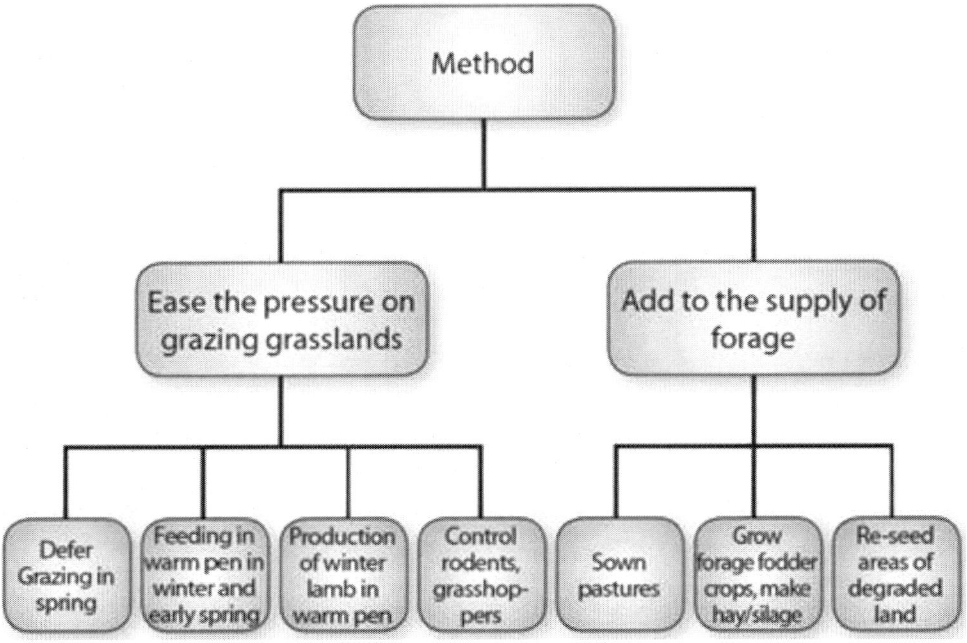


Figure 6. There are only two main options open to herders in western China.

(i) Fostering natural regeneration and rehabilitation of vegetation as seen for example, in the National project "Returning grazing land to grassland (RGLGL) referred to above. This is directed by ecological principles with less emphasis now on man-made landscapes that were the promoted under the 'ecological construction' phase.

(ii) Reform of institutions and changes in governance of rangelands combined with establishment of a stable input mechanism and public finance as a mainstay, improve the eco-compensation mechanism following the principles of endowing benefits on those who contribute to arresting and reversing degradation while being compensated by those who benefit. Improvement of the user rights and land tenure system to provide greater incentives to land users to adopt a land stewardship role secure in the knowledge that more secure tenure should allow the present and future generations to benefit from implementation of more sustainable practices.

(iii) Promote economic and social development in N and NW China (in particular), control population growth and migration to urban areas[8] to gradually reduce population pressure and improve living standards, Economic structure in affected areas will be 'optimized' and linkages between agriculture animal husbandry and forestry will be adjusted and localized processing of farm produce, including fiber and skins, meat etc. will be established. Emphasis will be on alleviating poverty and increasing incomes of land users.

(iv) Develop better inter-agency cooperation to reduce the 'silo' effect, avoid overlap and gaps and improve efficiency and eliminate conflicting policies being promoted by different bureaus. An incentive and penalty system to be set up and implemented to become part of the assessment of local governments' work. An improved scientific

[8] China's policy is speed up the urbanization and reduce population pressure on fragile areas.

support system to be established and proliferated to complete the monitoring and early warning network.

(v) Address the problem identified by Brown, Waldron and Longworth (2006) as 'Uniform Policies, Non uniform Impacts' As they observe

(vi) *"...policies aimed at rural, regional, industry and environmental development in intensive agricultural regions of China are not always, and indeed not often, suited to the extensive pastoral region which has vastly different ecological, social and economic problems " p.259. Despite then enormous differences the "Han approach" to agricultural policies has been transferred wholesale to pastoral regions".*

The condition of rangelands in China's pastoral areas is at a crossroads. Whether they take the route to rehabilitation and sustainable production or the road to further degradation hangs on three factors;

(I) The willingness of policy makers to embrace change, to make strategic decisions based on the best available scientific and socio-economic information and to facilitate change through new policy settings

(II) The amount the government is prepared to pay to repair the rangelands, develop sustainable rangeland production (albeit at a lower level than previously – many fewer herders), reduce dust storms and slow the movement of encroaching sand

(III)The rebirth of a sense of community and the embrace of a land stewardship ethic (Zhang, Borjigin and Zhang, 2007 and Zhang, 2006, Squires, 2012).

Whilst it is true that the situation in N and NW China can be summed up as "Too many people, too many livestock, too little feed' there is a clear need to tackle the causes of the rangeland degradation problem and not just deal with the consequences.

Most of the effort in the past has aimed at 'solving' minor problems such 'how to get more forage from each hectare' rather than deal with the underlying causes of low productivity such as insecure land tenure, unclear boundaries for the assigned grazing user rights, and lack of clear policy on how to balance livestock numbers and feed supply. Although it has been stated that the proximate cause of rangeland degradation is the imbalance between feed supply and livestock carrying capacity, the underlying cause is related to people -- the users of the land (Squires, 2009).

## ACKNOWLEDGMENT

The authors are grateful for the feedback and advice from Dr LM Ding at Lanzhou University and for the assistance of the anonymous referees.

## REFERENCES


Brown. C., S. Waldron and JW. Longworth, 2008. Sustainable Development in Western China: Managing People, Livestock and Grasslands in Pastoral Areas. Edward Elgar Publishing, Cheltenham, UK 294.p.

Dalintai, A. 2008. Rethinking overgrazing and strategies for its management in Inner Mongolia. In: Multifunctional Grassland in a Changing World -- Proc. International Grassland Congr. Vol 1, Paper 1676, Guandong People's Publishing House.

Dong, S. K., H. W. Gao, G. C. Xu, X. Y. Hou, R. J. Long, M. Y. Kang, and J. P. Lassoie, 2007. Farmer and professional attitudes to the large scale ban on livestock grazing of grasslands in China. *Environmental Conservation*, 34(3):246-254.

Halimova, N. 2012 Land Tenure Reform in Tajikistan: Implications for Land Stewardship and Social Sustainability: A Case Study. pp. 305-329 In: V. Squires (ed) 'Rangeland Stewardship in Central Asia: Balancing Livelihoods, Biodiversity Conservation and Land Protection'. Springer, Dordrecht.

Harris, R. B. 2010. Rangeland degradation on the Qinghai-Tibetan plateau: A review of the evidence of its magnitude and causes. *Journal of Arid Environments,* 74: 1-12.

Heshmati, GA. and V. Squires, (eds) 2013. 'Combating Desertification in Africa, Asia and the Middle East: Proven Practices'. Springer, Dordrecht.

Ho, P. 1998 Ownership and Control in Chinese Rangeland Management since Mao: A Case Study of the Free-rider Problem in Pastoral Areas in Ningxia P. R. China In: E. B Vermeer, W. L. Chong-Woerkom and F. Peike (eds) Between State and Private Interest: Chinese Rural Collectives and Voluntary Peasant Organisations. M.E., Sharp Publishers.

Hua, L. 2009 Application and Practice of Integrated Ecosystem Management (IEM) in Gansu GEF Pastoral Development Project. pp.227-238 In: Jiang Zehui (ed) Integrated Ecosystem Approach and Management. China Forestry Publishing House, Beijng.

Hua, LM. and VR. Squires. 2015 Managing China's pastoral lands: current problems and future prospects, *Land Use Policy*, 43 (2015) 129–137.

Hua, L.M. and D. Michalk. (2010) Herders income and expenditure: Perceptions and Expectations. pp. 233-253 In: Victor Squires Limin Hua, Degang Zhang and Guolin Li (eds) Towards Sustainable Use of Rangelands in North-west China. Springer Dordrecht.

Hua, LM., SW Yang, VR. Squires and GZ Wang 2015. An Alternative Rangeland Management Approach in an Agro-Pastoral Area in Western China Range Ecology and Management 68:109-118. doi.org/10.1016/ j.rama.2015.01.001

Hua, L. and D. Zhang. 2010. Engaging with Land Users: The First Steps on a Long Road pp. 333-356 In: Victor Squires (ed) Rangeland Stewardship in Central Asia: Balancing Livelihoods, Biodiversity Conservation and Land Protection. Springer, Dordrecht.

Kirychuck, B. and B. Fritz. (2010) Ecological control of rangeland degradation: Livestock Management pp. 61-79 In: Victor Squires, Limin Hua, Degang Zhang and Guolin Li (eds) Towards Sustainable Use of Rangelands in North-west China. Springer Dordrecht.

Li Q., D. Michalk, LX., Chen, P. Zhang and S. Tong. 2003 Analysis of constraints and management strategy for animal production in grasslands of northern China, *Grassland*, 11(2) 47-53 [In Chinese].

Li, X. LJ. Gao, G. Brierley, YM. Qiao, J. Zhang and YW. Yang. 2013. Rangeland degradation on the Qinghai-Tibet plateau: implications for rehabilitation, *Land Degradation and Development*, 24(1): 72-80.

Michalk, D., L. Hua, D. Kemp, R. Jones, T. Takshashi, J. Wu, Z. Nan, Z. Xu, and G. Han 2010. Redesigning Livestock Systems to Improve Household Income and Reduce Stocking Rates in China's Western Grasslands. pp.301-324 In: Victor Squires, Limin Hua, Degang Zhang and Guolin Li (eds) Towards Sustainable Use of Rangelands in North-west China. Springer Dordrecht.

Richard, C. (2005) Developing alternatives to resettlement for Pastoralists on the Tibetan Plateau. *Nomadic Peoples*, 9 1-2.

Shang, Z.H., M.J. Gibb, F. Leiber, M. Ismail, LM. Ding, X.S. Guo and RJ. Long, 2014. The sustainable development of grassland-livestock systems on the Tibetan plateau: problems, strategies and prospects. *Rangelands J.,* 36:267-296.

Sheehy, D.P., D. Miller, and DA Johnson, 2006. Transformation of traditional Pastoral Livestock Systems on the Tibetan Steppe. *Secheresse,* 17(1): 142-51.

Squires, VR. (ed) 2009. People in Rangelands. In: 'Range and Animal Sciences and Resources Management', Encyclopedia of Life Support Systems (EOLSS), Developed under the Auspices of the UNESCO, EOLSS Publishers, Oxford, UK.

Squires, VR. Lu Xinshi, Lu Qi, Wang Tao and Yang Youlin 2009. 'Rangeland Degradation and Recovery in China's Pastoral Lands', CABI, Wallingford, UK, 264 p.

Squires, V.R., Hua Limin, Zhang Degang, and Li Goulin 2010. 'Towards Sustainable Use of Rangelands in North West China'. Springer, Dordrecht 353p.

Squires, V. (ed) 2012. 'Rangeland Stewardship in Central Asia: Balancing Livelihoods, Biodiversity conservation and Land Protection'. Springer, Dordrecht p.458.

UNCCD, 2013. Economic and Social Impacts of Desertification, Land Degradation and Drought. White Paper 1, Prepared for the Second Scientific Conference, 9-12 April, 2013 Bonn, Germany 59p.

Wang, XH., Lu, C. Fang, J. and Y. Shen 2007. Implications for development of grain-for-green policy based on cropland suitability evaluation in desertification-affected north China. *Land Use Policy*, 24(2,) 417–424.

Williams, A., M. Wang, and M. Zhang, M. 2009. Land Tenure Arrangements, Property Rights and Institutional Arrangements in the cycle of Rangeland Degradation and Recovery pp.219-234 In: V. Squires, X. Lu, Q Lu, T. Wang and Y. Yang (eds) Rangeland Degradation and Recovery in China's Pastoral Lands CABI Wallingford.

Williams, D. 2002 'Beyond Great Walls: Environment, Identity and Development on the Chinese Grasslands of Inner Mongolia'. Stanford University Press, Stanford.

Zhang, MA., E. Borjigin, and H. Zhang. 2007. Mongolian nomadic culture and ecological culture: on ecological construction in the agro-pastoral mosaic zone in Northern China. *Ecological Economics,* 62 (1)19-26.

Zhang, Q. 2006. May they live with herds: Transformation of Mongolian pastoralism in Inner Mongolia. Masters Thesis, University of TromsØ, Norway 101p.

In: Rangeland Ecology, Management ...
Editor: Victor R. Squires

ISBN: 978-1-63482-504-7


*Chapter 3*

# SMALL HOLDERS, THE KEY TO SUCCESSFUL IMPLEMENTATION OF INTEGRATED ECOSYSTEM MANAGEMENT: A CASE STUDY FROM NORTH WEST CHINA'S PASTORAL ZONE

***Hua Limin and Victor R. Squires***
College of Grassland Science, Gansu Agricultural University,
Lanzhou, China

## ABSTRACT

Rangeland degradation is widespread in western China. The lack of a sound science base, coupled with the cumbersome administrative system, means that implementing more sustainable approaches face many challenges in China. In 2003, GEF initiated a project under Operational Program 12 (OP12) to support pastoral development in western China. The project focused at the village level and worked with the small holders to implement an integrated approach, which combined social science and natural science, rangeland management and livestock production in an effort to arrest and reverse rangeland degradation. The challenge was to meet the needs of the users while at same time identify a package of interventions that would improve livelihoods while restoring rangeland productivity. This chapter gives a brief overview of the project, shares the experiences during the 5 years of project implementation in 9 counties in the Hexi Corridor of Gansu Province in the arid region of north-west China's pastoral zone and lists the lessons learned in the belief that these will be of value to people in other developing counties.

**Keywords**: GEF project, small householders, integrated approach, rangeland management, NW China

# 1. INTRODUCTION

Land degradation and biodiversity loss are the most critical issues relating to the ecological environment in the west of China and they are the main causes for poverty and constraints for economic development (Squires et al., 2010; Hua and Michalk, 2010; Waldron, et al., 2011). The Chinese government has paid much attention to these issues, and significant achievements have been achieved in land degradation control through long-term efforts, especially under the auspices of the Great Western Development Strategy (Wang and Wei 2003; Cao, 2011). However, although the related governmental departments in terms of traditional or modern approaches have made active efforts for land degradation control and biodiversity conservation, the outcomes are limited because these efforts are from a single department/bureau and the policies are not sustainable (Zhang, et al., 2008; Feng, et al., 2005, Kemp, et al., 2011) and land degradation became very severe (Lu et al., 2005)

Degradation of rangeland (commonly called grassland in China) is a complex problem in China (Squires et al., 2010; Hua and Michalk, 2010; Li, et al., 2007). The principal pattern of rangeland-based livestock production systems in China is a householder family running a small number of livestock, which is different from the commercialized operating pattern (Longworth and Williamson 1993; Banks et al., 2003). At first glance it appears that rangeland degradation is due to an unbalanced feed and animal relationship caused by over grazing, but the root causes lie elsewhere (Squires, 2009). Many problems, apart from those over which the land user has no control (weather, government policy, markets) that exist in the development of rangeland based livestock systems are closely related to the land user. So capacity building of farmers/herders is one of the key pre-conditions for sustainable rangeland development (see Figure 1).

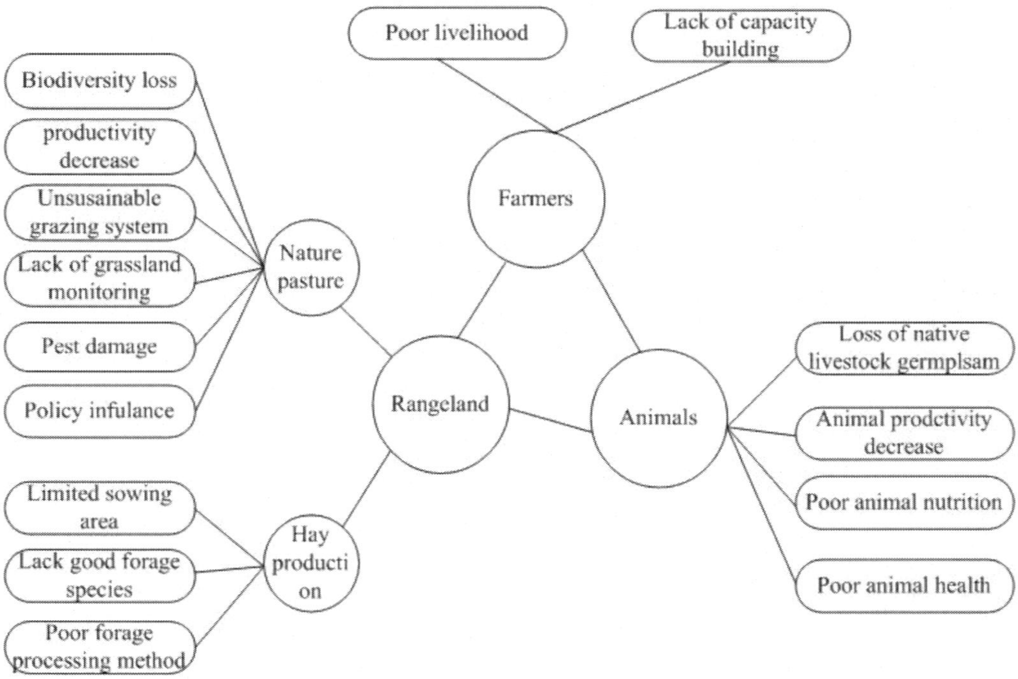


Figure 1. Major problems existing in pastoral development in GEF project areas of Gansu.

In 2003, GEF (Global Environment Facility) initiated a project under operational program 12 (OP12) to support the pastoral development in Gansu and Xinjiang with 10.5 million US Dollars and it is a valuable attempt for resolving the increasing land degradation in China (World Bank, 2003). The project is the first large scale operational project of the China-GEF strategy partnership on land degradation in dryland ecosystems under the GEF/OP12. It is funded by the GEF and linked to the World Bank Gansu and Xinjiang Pastoral Development Project. The project is consistent with three operational programs (biological diversity, climate change, and land degradation) of the six Global Environment Facility (GEF) focal areas (biological diversity, climate change, international waters, ozone layer depletion, land degradation, and persistent organic pollutants) and with Chinese Government's strategy on the Convention on Biological Diversity (CBD) and the Convention to Combat Desertification (CCD). The project responds to Conference of the Parties (COP) to the CBD guidance in that it promotes capacity building, conservation and sustainable use of natural resources through adaptive management of grassland landscapes, and supports the objectives of above-mentioned international conventions.

The project had its foundation in Integrated Ecosystem Management (IEM) defined as a holistic approach to address the linkages between ecosystem functions and services (such as carbon uptake and storage, climatic stabilization and watershed protection, and medicinal products) and human social, economic and production systems" (Jiang, 2007). As such the project in China has great relevance to Central Asian, African and other developing countries both in the demonstrated outcomes but, more importantly, the benefits of an approach that delivers a package of interventions to deal with what are complex biophysical and socio-economic problems of those faced with accelerating land degradation, lower household incomes and reductions in their quality of life.

## 2. Brief Overview of the GEF Project as Implemented in Gansu

### 2.1. Project Aim and Objectives

The project aims to mitigate land degradation, conserve globally important biodiversity, and enhance carbon sequestration. The global environmental objective of the project was to maintain and nurture rangeland ecosystems to enhance global environmental benefits. In the interest of brevity the 10 Objectives are dealt with in section below.

### 2.2. Project Area

The project areas covering nine counties in Gansu (Figure 2) are mainly located in the Qilian Mountains and the rangeland in the Qilian Mountains is the headwaters of the Shiyang River, Hei River, Shule River basins which comprise totally 56 tributaries (Figure 3). These inland rivers are the only water resource of the oasis agriculture zone in Hexi Corridor (Bao and Fang, 2007), that runs westward from Lanzhou for over 1000 km. So, the success of oasis agriculture and livelihood of local people directly depend on the ecological status of rangeland in the Qilian Mountains. The rangeland types in Qilian Mountains, the piedmonts and adjacent plains are diverse with rich biodiversity, typical of cold and arid high altitude desert, alpine meadow, temperate steppe and sand deserts (Table 1). The project basically covered the major types of land utilization in the north-western areas of China; including oasis irrigated agriculture, rainfed farming, and various typical mountainous steppe and desert steppe.

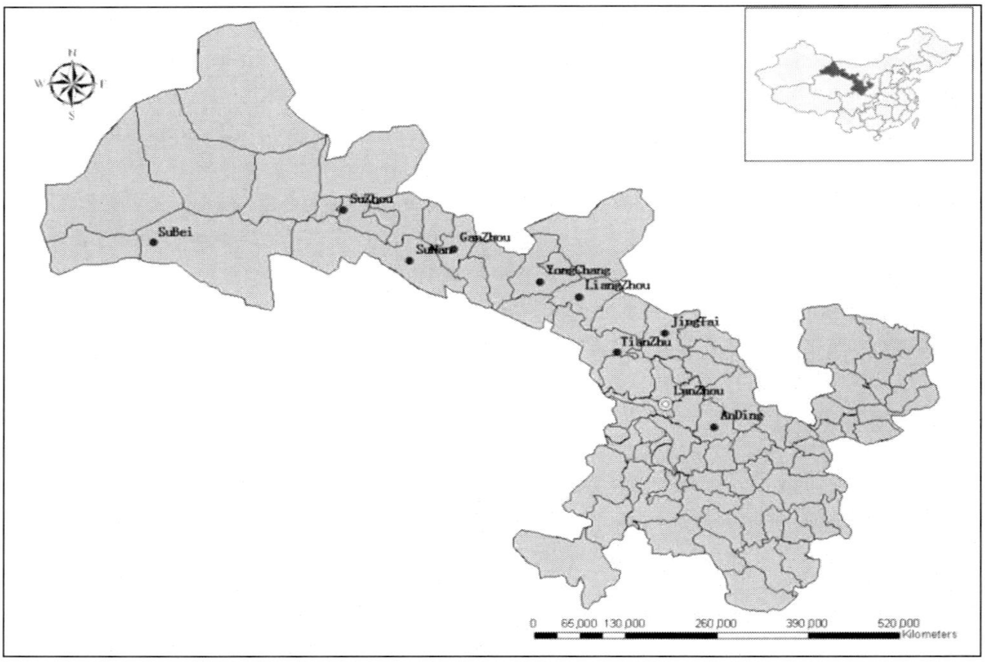


Note: • represents a GEF project implementation county.
Figure 2. Location of GEF project counties in Gansu.

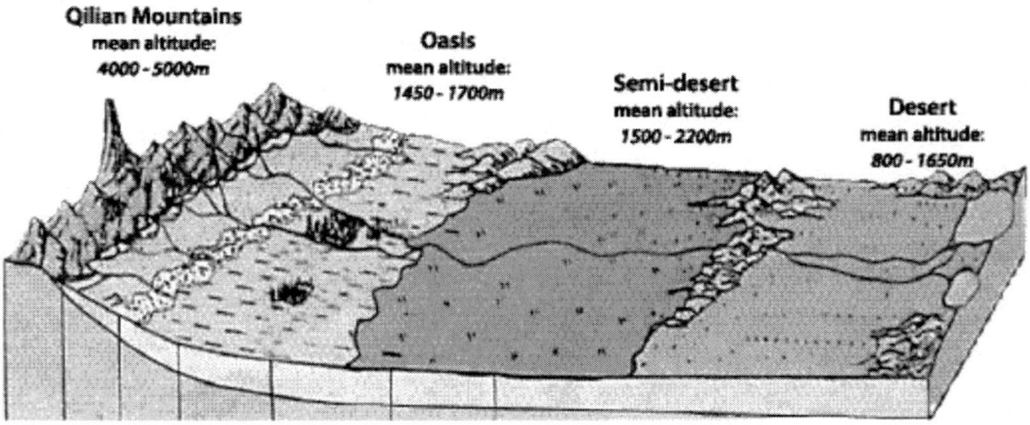


Figure 3. The landscape of the Hei River basin. The Hei River is typical of the three river basins that flow northward from the Qilian mountain chain.

**Table 1. Forage yield as an indicator of rangeland degradation in selected counties in Gansu Province, western China**

| County | Type of grassland | Average Yield* | Average change in yields from 1985 to 2006 |
|---|---|---|---|
| Anding | Temperate Meadow Steppe | 1020 | -34% |
| Jingtai | Temperate Steppe | 443 | -61% |
| | Alpine Meadow | 1001 | -39% |
| Liangzhou | Temperate Steppe | 602 | -21% |
| Yongchang | Temperate Steppe | 531 | -55% |
| Ganzhou | Temperate Desert | 440 | 27% |
| Sunan | Alpine Meadow | 1004 | -72% |
| | Alpine Shrub- Meadow | 800 | -14% |
| Suzhou | Lowland Meadow | 1665 | -61% |
| Subei | Temperate Desert | 659 | -61% |
| | Alpine Meadow | 438 | -51% |
| Tianzhu | Shrub- Meadow | 536 | -31% |

* Yield as Dry Matter (kg/hectare) (Data source: Gansu Grassland General Workstation).

## 2.3. Main Project Beneficiaries

The Qilian Mountainous area is one of the areas in Gansu where the minority nationalities predominate. Pastoralism is their backbone industry that they rely on for their existence. Land degradation has already impacted negatively upon those minorities' livelihood, social stability, even the minority's culture that is in danger of being lost. The project beneficiary groups include Hui, Mongolian, Yugur and Zang (Tibetan) etc. altogether 1270 people of 225 households.

## 2.4. Project Activities

Table 2 is a summary of the key components and interventions that were developed under three approaches. The project-targeted activities that increased forage yield and enhanced rangeland stability and long term productivity. Rangeland monitoring was established at several sites within each project area to provide a basis for assessing the effectiveness of management interventions.

In addition to technical training, the project carried out an environmental education (EE) program for pupils in elementary school of the project area (Zhao, Hua and Squires 2010). The activities involved an EE course in the classroom and EE practice in the field. The emphasis was on teaching children to care for the environmental matters in their neighborhood, and influence other family members to pay attention to the local environmental matters.

Due to the importance of the consultant service in the GEF project (consulting service cost account for 40% of the total investment) and according to the technical problems and the farmer and herder's requirements; a total of nineteen applied researches and technical services were designed and conducted in the project. This project includes three aspects: resource investigation, socio-economic research, and technical assistance. Figure 4 is a summary of the topics of the applied research projects and their interrelatedness to the overall objectives.

**Table 2. Three strategic approaches in GEF project in NW China**

| Approach 1 | Approach 2 | Approach 3 | |
|---|---|---|---|
| Livestock and rangeland development | Institutional/policy development | Diversification | |
| 1.1 Forage production<br>1.1.1 Sow high quality forage<br>1.1.2 Improve forage processing and conservation method | 2.1 Land tenure reform<br>2.1.1 Trial of alternatives | 3.1 Non-livestock agricultural production | 3.2 Non-agricultural production |
| 1.2 Breed improvement<br>1.2.1 Preserve local breeds<br>1.2.2 Crossbreeding | 2.2 Policy adjustment | 3.1.1 Alternative energy<br>- Solar,<br>- biogas | 3.2.1 Review employment opportunities regional and district plans |
| 1.3 Pasture management<br>1.3.1 Rest rotation grazing<br>1.3.2 Deferred grazing<br>1.3.3 Fencing<br>1.3.4 Biodiversity<br>1.3.5 Seed collection and reseeding | 2.3 Capacity building FFS* for improved husbandry<br>2.3.1 Farmer/community organisations | | 3.2.2 Other investment<br>3.2.3 Business opportunities |
| 1.4 Feeding & nutrition<br>1.4.1 Ration formulation<br>1.4.2 Mineral brick | 2.4 Rural finance | | 3.3.3 Capacity building<br>3.3.3.1 Vocational training<br>3.3.3.2 Study tour both inland and abroad |
| 1.5 Warm pens for winter | 2.5 Environmental education | | 3.4 Consultant services |
| 1.6 Water supply<br>1.6.1 wetland management<br>1.6.2 Built drinking spot | 2.6 Training | | 3.5 Applied research |
| Pest control<br>Rodents<br>Grasshoppers.<br>Toxic plants control | 2.7 Markets<br>2.7.1 Livestock Auctions | | |

Note: 'FFS=Farmer Field Schools'

# 3. Outcome of Project Development Objectives

## 3.1. Developed the Village-based Participatory Grassland Resource Management Plan

The farmers/herders are the main users of grassland and their knowledge and actions are very important to grassland management. The project supports all GEF villages to establish a grassland management group and establish a community grassland management regulation (CGMR) based on the group members' discussion. The group members should consider the practicality of changes taking account of vulnerable groups. Therefore, the group consists of village governor, villagers and women, etc. The CGMR has a particular target due to stakeholders' contribution. For instance, in pure grazing areas, such as Sunan, and Subei counties, the project supports the local community to establish the grassland management group. The group members made a detailed grazing management regulation, which aims at defining the onset and cessation of grazing on different seasonal pasture and punishment for herders who break the regulation. Compared with the former grassland management regulation made by the local community, the new CGMR has several differences: (i) the CGMR have been made by the villagers themselves based on deep discussion and negotiation, which is democratic management. (ii) the person who metes out punishment is from the group, (iii) the other members trust their neighbor and appreciate the benefits of the new arrangements. All of these factors reduce community conflict. Based on the CGMR, the project also develop the participatory grassland management plan through participatory *PRA tools*, which will fit the principle of sustainable grassland utilization, be feasible and be accepted by local farmers. The *annual project implementation plan* (PIP) of each GEF project sites accorded with grassland management plan.

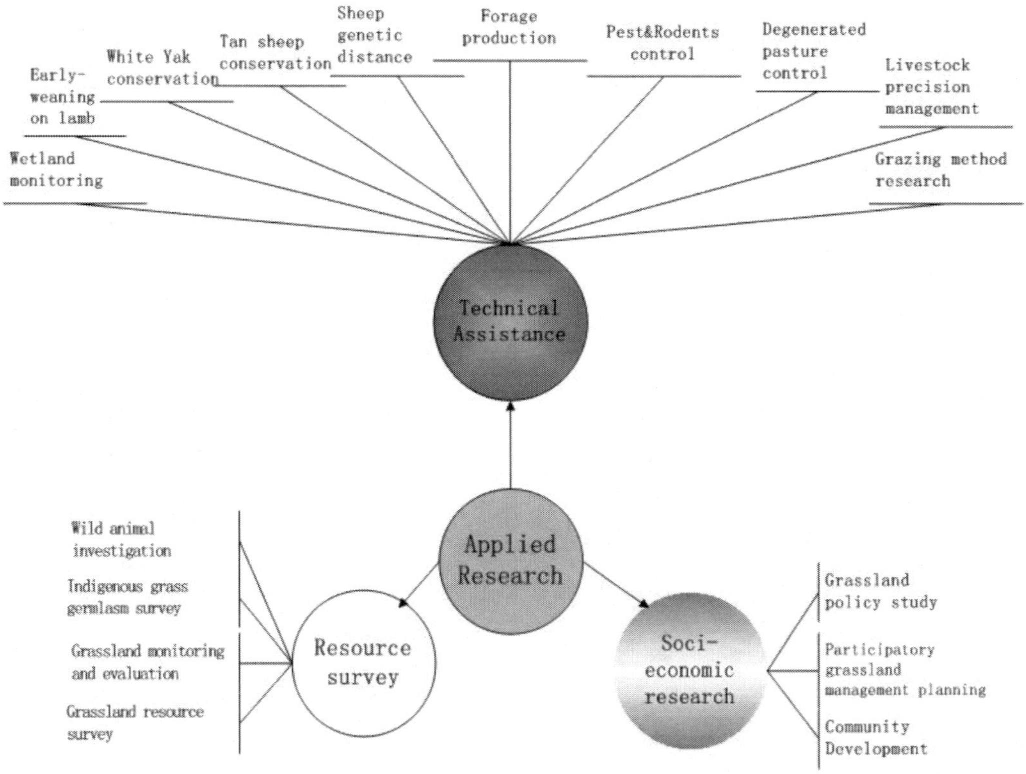


Figure 4. Consultant services and their role in GEF project in Gansu.

## 3.2. Developed the Grassland Survey

The latest Province-wide grassland survey (covering $1790.42\times10^4$ $hm^2$) was carried out in 1980s. At present, there is no updated information of grassland degradation or rehabilitation, such as area, location, grade, etc. on which the government can make the policy for controlling grassland degradation, regional planning or extending new technologies for grassland management. The GEF project invited a consultant team to carry out grassland survey in eight GEF counties. The approach combined remote sensing and investigation in field. After the hard work over three years, the consultant team completed the grassland survey reports, grassland resource map and grassland productivity grade map with scale of 1:250,000 (Figures 5-7), and the establishment of a grassland resource database. The work can effectively reflect the change of grassland utilization and degradation, which can guide the grassland management for governments and researchers.

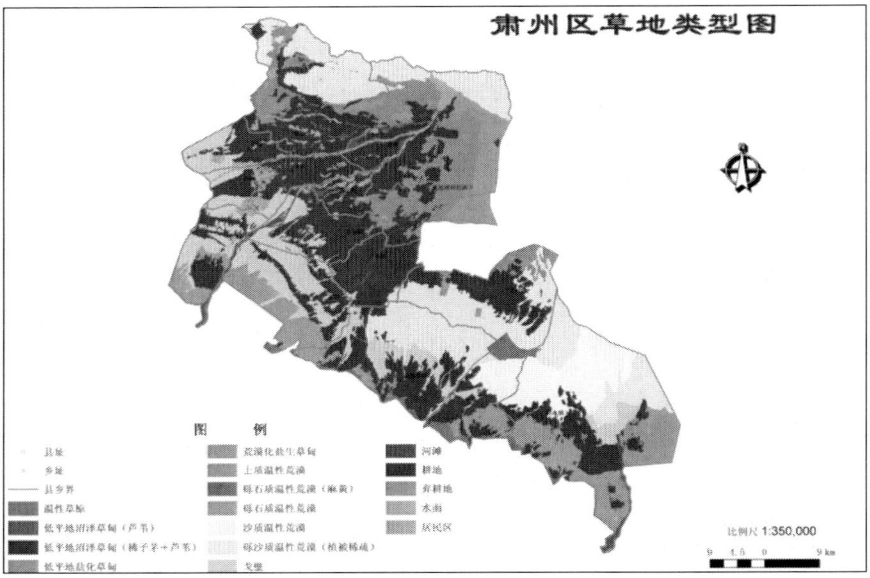


Figure 5. The grassland types in Suzhou District, Gansu, China. Altitude and aspect play a major role. Higher N facing slopes are cold.

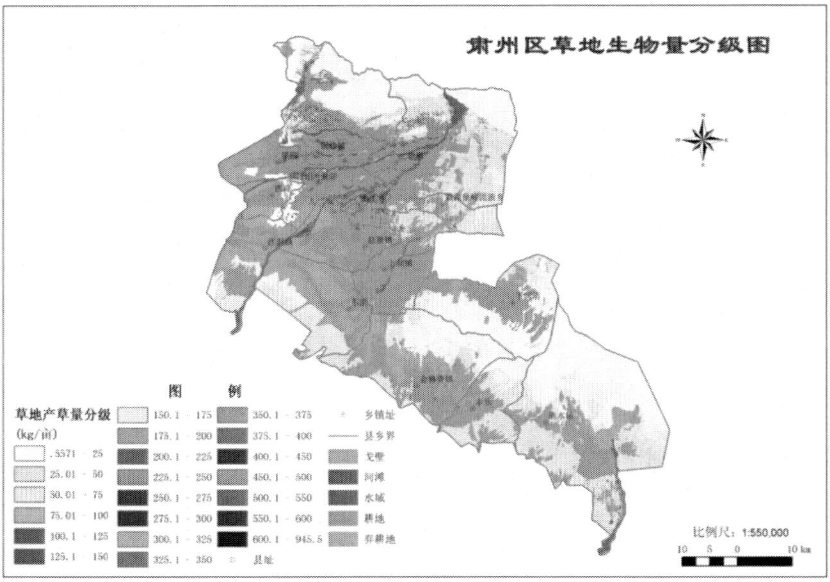


Figure 6. The classification map of grassland productivity in Suzhou District, Gansu, China.

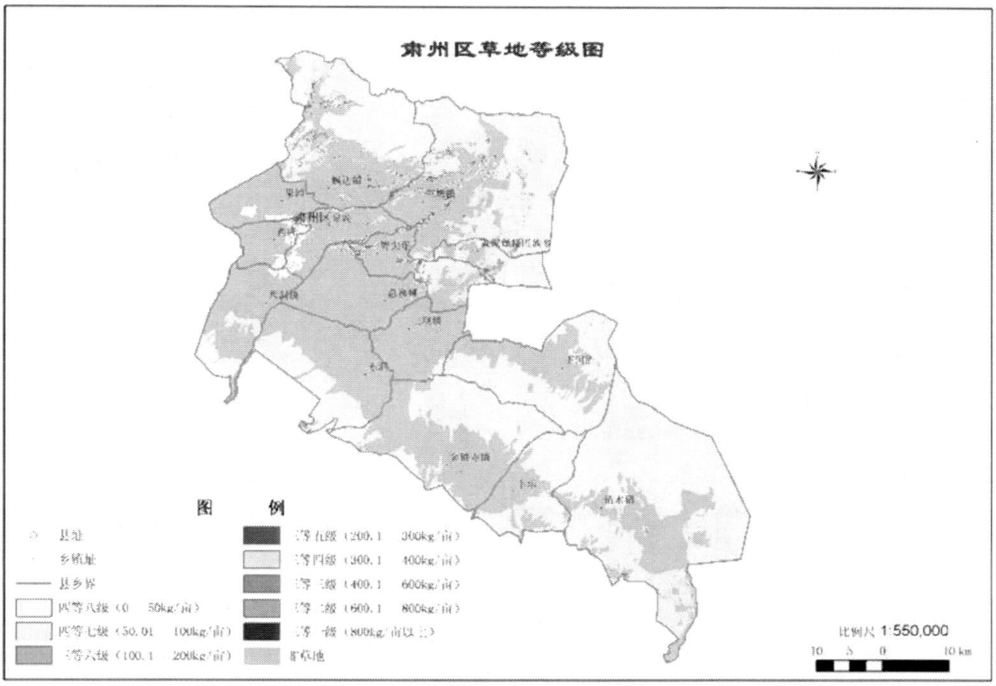


Figure 7. The grassland grade map in Suzhou District, Gansu, China.

## 3.3. Developed the Forage Production and Increase of Forage Supply that Benefits Mitigation of Land Degradation

The main proximate cause of natural grassland degradation is imbalance between grass and animal, especially where the reality of the unalterably high grazing pressure must be faced in the short to medium term. Increasing the forage supply is the major measure to solve the contradiction between grass and animal. For the various land types in the GEF project area, the activities for increasing forage production were purposefully adopted. In cropping areas such as Anding, Jingtai, Suzhou etc. with a larger area of cropland, enlarging the area devoted to forage production and introducing good quality forages are the approaches for increasing forage yield and quality. In pastoral areas, Sunan and Subei etc, the main activities are introducing good forage species to increase yield on the limited areas where soils are deeper and supplementary irrigation is available.

In the project districts with significant areas of artificial grassland, a mower and grass chopper were purchased and the silo-pit was built for farmers/herders to change the conventional pattern of forage processing, increase forage quality, and emphatically solve the problem of forage deficiency in winter and spring. By these measures, the forage productivity has been improved. In Anding District, the area planted to sanfoin (*Onybrychis* spp.) by the project households increased from a modest 20.7 ha before the project implementation to 120 ha by 2010. Comparisons of the beneficiaries vs non-beneficiaries, the planting forage area per householder was 14.13 ha and 5.75 ha respectively. The cropping structure had an obvious change and underwent rationalization (Figure 8).

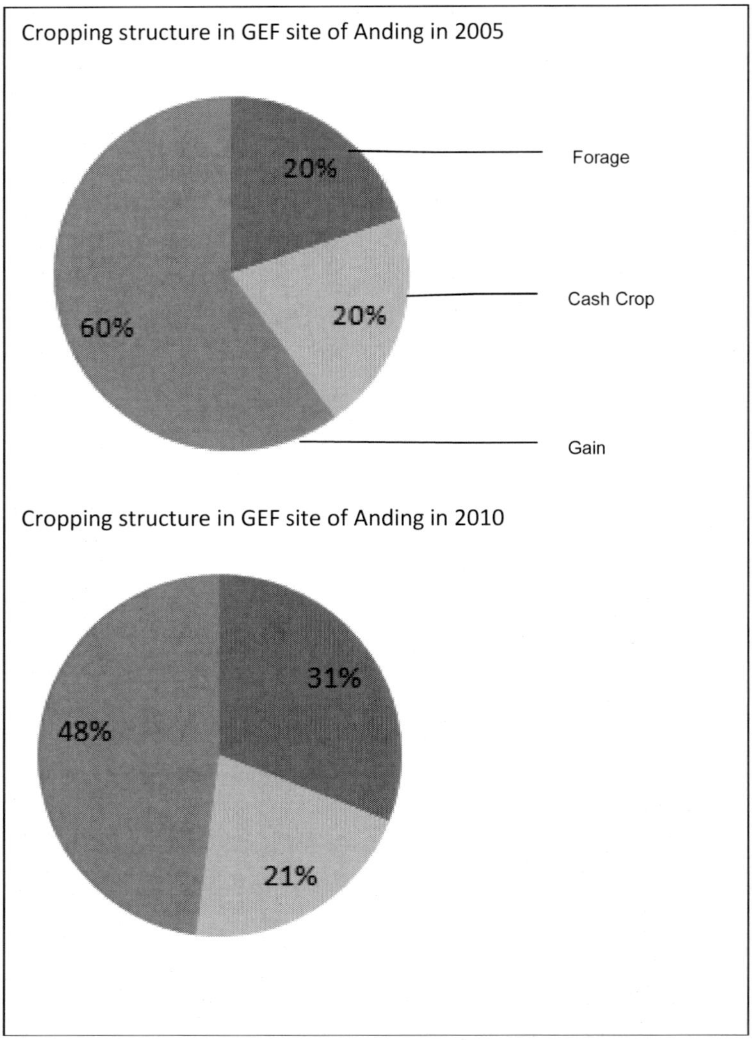


Figure 8. The change of cropping structure in Anding, Gansu.

Due to an expansion of the area sown to forage, combined with pasture fencing, the soil and water erosion has been effectively controlled. The soil loss was 5640 ton km-$^2$ p.a.before project implementation but, after fencing of grassland on the top of hill as well as planting forage on slope, soil loss decreased to 1666 ton per km$^2$.p.a. The coverage, height and biomass of natural pasture increased from 86%, 8.2cm, 3225 kg/hm$^2$ to 97%, 14.6cm, 5355 kg/hm$^2$.

## 3.4. Introduction of Livestock Breed and Focus on Quality Instead of Quantity

To solve the issue of feed balance, the available approach is to improve livestock productivity via a reduction in animal number. Where it is possible the development of forage production is fostered. It is not always possible, especially in the Tibetan Plateau, western

China. The development of forage production on a large scale is not possible because of its harsh environmental conditions. The feasible approach is to reduce the livestock number on the pasture and increase the farmers' income through improvement of livestock product. In Sunan, the project supported the herders to introduce the wild bull yak with 1/2 consanguinity to improve body size of their yaks. The average birth weight of improved offspring was 2.1 kg heavier than the ordinary yak calf. The body weight at six months and one year were heavier than the ordinary yak calf (8.5 kg and 13.9 kg respectively). The total yak number was reduced by 72 heads in the village because high quality improved breed. This saved labor and increased production (Table 3).

**Table 3. Improvement effect of Tadong bull yak on native yak**

| Index | Improved calves | Local calves | Changes |
|---|---|---|---|
| Birth weight (kg) | 13.3 | 11.21 | 18.64% |
| Weaning weight (kg) | 67.65 | 59.2 | 14.27% |
| Weight of 12 months old (kg) | 82.28 | 68.39 | 20.31% |
| Weight of 18months old (kg) | 151.22 | 121.56 | 24.40% |
| Average carcass weight (kg) | 72.85 | 57.25 | 27.25% |
| Meat weight (kg) | 59.65 | 41.5 | 43.73% |
| Dressing percentage (%) | 49.56 | 47.31 | 4.76% |
| Pure meat percentage (%) | 40.58 | 34.30 | 18.31% |

In another project area, Ganzhou, the project beneficiaries introduced male Erdos goats, whose cashmere yield is high, to crossbreed with local goats. The yield of cashmere increased from 0.15 kg to 0.3 kg per goat and the farmers' income increased too. As a result of training, the project beneficiaries understood that it was important to protect grassland, and to focus on quality and not quantity. The grazing pressure has been reduced year by year. The over-grazing rate decreased from 161% in 2005 to 56% in 2009. However, non-beneficiaries households increased livestock numbers in order to ensure normal livelihood and production needs. This led to accelerated land degradation.

## 3.5. Developed the Warm Sheds for Increasing the Farmer's Income and Reducing the Grazing Pressure on Nature Pasture

Combined with grazing bans and rest grazing in the project area, the project built the warm sheds for farmers and encouraged them to practice indoor fattening for sheep or cattle. In Jingtai County, thirty householders were selected as recipients of warm sheds each of about 75 $m^2$ and seed for forage planting, and a silage pit as well. Compared to before and after implementation, the survival rate of lambs increased from 80% to 97%, the lamb body weight of six months old in the warm sheds was 30.8 kg per head with average weight gain 14.5 kg per head. However, the lamb body weight of six months old grazing outdoors was only 26.86 kg with average weight gain 9.82 kg per head. The net income of lamb in the warm sheds was sixty CNY Yuan higher than 34.6 CNY Yuan of lamb that relied on grazing. Due to the extension of warm sheds and forage production, the 2200 sheep of this village shifted to warm sheds from winter pastures, which reduced the grazing pressure on the grassland. Body weight loss in the winter was minimized, lamb birth weights and survival were higher too.

## 3.6. Explored the Reforms of Grassland User Rights (GUR)

It is reported that the most serious grassland degradation occurs in the agro-pastoral transition zone, where the grassland user rights in the area belong to local community and the farmers use the pasture under a system of open access grazing, leading to excessive grazing pressure and land degradation. For solving the issues, the project tried to conduct the reforms of GUR in Yongchang County, a GEF project area. In Yongchang, the pilot village has a small area of grassland and a high population density, which means the grassland could not be divided according to each household entitlement according to Grassland Contract Responsibility System. At the time of initial PRA in this village, the farmers suggested that they wanted to use and manage the pasture based on a linked householders system. After several discussions among farmers, villagers, officials and project coordinators, the plan was finalized. About 15 householders who lived in the village, linked together and formed the group to contract and manage part of the community-owned grassland. The farmers designed the fence area, stock access routes and watering points, etc. and fenced the pasture by themselves. Based on local farmer's discussion, 1127.47 $hm^2$ of winter pasture in 2006 was divided into the southern and northern community with 8 groups, which had been contracted by 143 householders. Each group selected a leader and worked out the corresponding management measures, basically to reduce the overgrazing livestock of some farmers and monitor each other for using the pasture for long term sustainability. The entry time after green up and end of season exit time was decided as well. The original grassland belonging to community was an open access resource leading to a "tragedy of the commons' scenario. Although grassland area of contracted management was small, which accounted for only 13% of 8672.82 $hm^2$ grassland in whole village, the effect was obvious. Firstly, the consciousness of farmers was changed. The awareness of grassland protection and sustainable development was strengthened. Especially, as for farmers carrying out contracted management of grassland, they not only maintained their fenced enclosure but also supervised each other. A grassland management system was put it into effect and sheep numbers were reduced by 3000 Sheep Units (SU) lowering the stocking pressure on the grassland. There were 5025 SU in the northern community and 3956 SU in the southern community before contract and 2018 SU in northern community and 1794 SU in the southern community after contract (Figure 9). The vegetation of natural grassland was restored. Since the reforms, the time of entry in summer pastures and dates of return to winter pasture was regulated after all farmers had signed the contract for grassland management. The system for seasonal utilization of grassland was put into effect. In 2008, the village was funded by a national project "Returning grazing land to grassland" to fence grassland to allow rest-rotation grazing and seasonal grazing bans to be introduced.

The farmers thought the GEF project mode of grassland management was suitable for the local situation and strongly recommended its adoption in the National project (Hua et al., 2015).

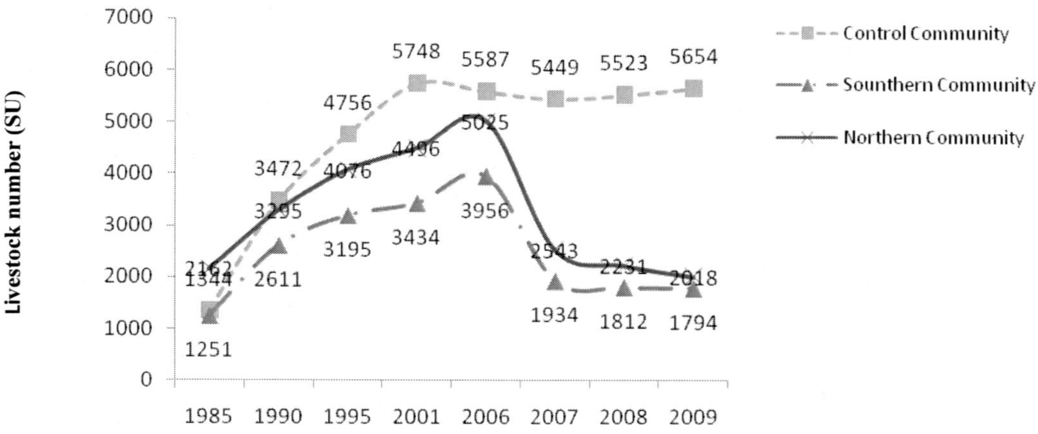


Figure 9. The change of livestock number in three villages. The Control group continued to use the communal grazing land on a year-round basis.

## 3.7. Developed the Biodiversity Survey

In order to reflect the status of biodiversity resource in GEF project area, the project conducted a biodiversity survey in eight counties. This involved the investigation of the plant resources, including kinds and distribution of the available forage species, native grass species and noxious plants, the animal resource that include breeds, number, distribution of livestock, wildlife and pest species on the natural grassland. Based on the survey, the reports and the database of biodiversity resource has been completed in the GEF project area. Additionally, based on a technicians' experience of Jingtai County, under the project coordinator's participation and assistance, with the GEF project financing "Higher Plant Resource in Jingtai County" was published in 2008. The book not only records the status of plant resource in detail in Jingtai County but also established a good foundation for conservation of biodiversity.

Due to the rich wildlife in Sunan and Subei, the Gansu Provincial Grassland Supervision Station was employed to undertake the wildlife resources investigation in the two counties. The investigation has involved the number, distribution and status of endangered species, etc. A report incorporating a database and a 1: 250,000 wildlife resources distribution map were published. The outcome can be used by the local government and researchers to make policy for wildlife management and to study the wildlife.

## 3.8. Collection of Indigenous Grass Seeds for Preserving Grass Germplasm

At the outset of the project implementation, an area of natural grassland was fenced for collecting indigenous grass seeds in some project counties according to the World Bank/GEF Project Assessment Document (PAD). As the Project developed it became clear that it was not necessary to develop the collection and reseeding of the indigenous grass seeds in *each* of the project counties. This was because the causes of grassland degradation in each county was different, and the grassland degradation in some counties did not have to be improved by reseeding. The collection of indigenous grass seeds was very difficult because of its uneven maturity and shattering – problems that were not be solved by simply training farmers and herders in collection techniques. Therefore, the area for the collection and seed-increase of indigenous grass seeds was established only in Yongchang and Sunan counties. The program has collected indigenous grass seeds of 68 species, 50 genera and 19 families, and established the propagation base, including 1.5 ha of updating and conservation nurseries, 10 ha of observation and domestication nurseries and 32 ha of quality seed increase fields, all of which are used for the purpose of conserving, domesticating, observing, propagating, updating, producing and utilizing indigenous grass seeds. The database of grass germplasm has been constructed for realizing the information sharing.

## 3.9. Developed the Native Livestock Conservation for Agro-Biodiversity

In Gansu Province, Tianzhu White Yak has unique genetic value (Figure 10). Over time, cross mating resulted in mixed white Yak and black yak and, the pedigree and generation degree are not clear. So that the white yak population is degraded even facing the possibility of disappearing. The project supported establishing the core herd of white yaks of pure or near pure ancestry based on the different type of grassland and herd's regional trait to breed good calves. The aim was conservation of white yaks living in their original habitat. We combined bio-technology; and genetic improvement through artificial insemination (5000 straws of frozen semen were made) to increase the number of calves and broaden the base of pure or near-pure white yaks. After five years implementation, the six core herds with 312 yaks were established. The number of white yak belong to Class 1 and Class 2 accounted for 45% and 56% respectively, which increased by 14% and 16% compared with that before the project implementation. 600 yak calves were selected for breeding. 16-month-old individual live weight and wool weight averaged 87 kg and 1.39 kg respectively, which increased by 21.06kg and 0.19kg respectively. The adult yak productivity was increased obviously (Table 4).

**Table 4. Comparisons on productivity of white yak before and after GEF project implementation**

| Index | In 2005 | In 2010 | Changes |
|---|---|---|---|
| Number of White yak (ten thousands heads) | 3.94 | 4.12 | 0.18 |
| Live weight of calf in 16 months old (kg) | 65.94 | 87 | 21.06 |
| Cashmere weight per calf in 16 months old (kg) | 1.2 | 1.39 | 0.19 |
| Survival rate of adult yak (%) | 94 | 98.5 | 4.5 |
| Survival rate of reproduction for calves (%) | 63.1 | 74.2 | 11.1 |
| Rate of yak for sale (%) | 14 | 21 | 7 |
| Average carcass weight of adult yaks (kg) | 130 | 140.5 | 10.5 |
| Average wool yield per yak (kg) | 1.5 | 1.93 | 0.43 |
| Average cashmere weight per yak (kg) | 0.8 | 1.1 | 0.3 |
| Average daily milk yield per yak (kg) | 1.5 | 2.02 | 0.52 |

A

B

Figure 10. A. White yaks are less common than black yaks. In Tianzhu county they have been prized and bred for a long time. B. In more recent times there has been mixing of the gene pools as the black and white calf indicates.

The 34 herds with 1,800 yaks for selected for grading-up (Lang, Wang and Squires, 2010). The effective combination of living livestock and bio-technical conservation was realized, and the database for the breed conservation of White Yak was established to service for the breed conservation. Based on the project activities, under the GEF finance, the book "The Technology of White Yak Conservation" has been published.

## 3.10. Developed Environmental Education

While conducting the training for farmer and herder, the project coordinator found that the children education could have influence on the adult education. Therefore, a team was invited to conduct the environment education for the children at the schools within the project area. After much discussion and trial and error including investigation, evaluation, discussion, analysis and school practice etc., the Environment Education Handbook for pupil and the EE manual for teachers were developed for the experimental school in the each of the 8 project counties, a set of the tutorial book and courseware (VCD) were developed for the teacher, and various activities about the environment education were developed including, teaching in classroom, practice outdoor, the pupils environment protection work such as the composition, painting, handicraft products, and collecting articles. The EE activities brought on a huge backwash in school to stimulate the children's enthusiasm of loving environment and protecting environment, by the activity of small hand in big hand, the children's environment exerted the influence on the parent's awareness in protection of environment. (Zhao, Hua and Squires, 2010)

## DISCUSSION

Most farms in the pastoral area in China are small scale, family operations and are very diverse (Longworth and Williamson, 1993; Sheehy, et al., 2006; Wang, et al., 2010). Many households rely on access to communal grazing lands under the control of the local government. The farmers who rely on their farm are also the decision-makers related to grassland management, which means that farmers are left to assess the health and sustainability of rangeland. Government policy can also influence the approach of rangeland utilization through subsidies, grazing bans, fencing, etc. However, the farmers usually make a plan of rangeland management based on their knowledge and experiences. Agricultural extension services are poor or non-existent in the pastoral area in China. So far, whenever Chinese government or foreign support projects have extended new policies, improved infrastructure, or the utilization of new or modified technology or management options the improved technology is rarely adopted. The reasons were that approaches copied from other counties or from foreign donor-funded projects neglected the production model of small farmers, which is the key of sustainable system in developing countries. An understanding of their livelihood systems and careful analysis within the constraints of these systems is necessary to help change agents, infrastructure managers and policy makers tailor changes that would be relevant and favorable to the larger segment of the population.

Since 2004, the central government has issued a policy on defragmentation of land to solve the problem of small production system and poverty profile of farmers. In the pastoral area, the local government encourages farmers and herders to establish Farmers/Herders Associations to manage larger areas of rangeland as single unit to take advantage of spatial and temporal variability in forage and water resources and improve the efficiency of the grazing system. It is still being worked out how best to do this but one thing is clear from experience that the model of each household being operated as separate production unit is no longer viable in the market economy (Hua and Squires, 2015).

## CONCLUSION

1. To solve grassland degradation there is need to apply the "integrated package of activities", which not only involves the increase of the grassland productivity and the decrease of the grazing pressure, but also the problem of herder's livelihood and community development. The GEF project carried out the packaging of activities of

linked interventions. This was central to the success of the project, in particular when combined with capacity building at all levels from township to Province.

2. The participatory approach is the basis of project implementation. The application of participatory approach made the project decision-making more democratic and the project activities further suited to the local situation, especially the participatory M&E (monitoring and evaluation) was more scientific and helped in the regulation of project activities.
3. Capacity building is the guarantee of the long-term impact of the project. The implementation of the GEF project supplied a lot of training opportunities to the project managers, technicians and farmers at various levels. The training chance was not only the enhancement of management and technology transfer, but more important, was that the mind-set change of the people will have a long-term impact.
4. The impact on policy formulation and implementation is longer lasting. China's grassland degradation is a very complicated problem, which is not only the issue of assemblage of various techniques. The more important is the impact of policy, such as the land policy, economic policy in pastoral area, and multiple sectors cooperation. Therefore, to solve the contradiction of grassland degradation, the first thing that must be solved is the restriction imposed by the policy framework. Unless policy is modified the other technical measures cannot play an effective role. The reform of grassland user right based on the community was executed in the project, which was an attempt on the reformation of grassland user right in the agro-pastoral area, for the experiment and demonstration on the improvement of grassland management on the policy level (Hua et al., 2015b)
5. Pay attention to production pattern of small farmers. The principal part of China's grassland animal husbandry consists of the pattern of the family operation with the household as the production unit, which is not as same as the operation pattern in the commercial pasture in advanced counties. Based on this point, the Gansu GEF project was mostly focused on the farmer and herder level. More attention was given to the farmer/herder livelihood and to the replicability of the successful interventions through the extension service. For example, compound silo-pit and grass chopper should be of small type and suitable for farmer to use and the technology easily replicated.

## ACKNOWLEDGMENTS

The authors are grateful to the many people in the 8 counties who have contributed to the outcomes of this project. The PPMO in Lanzhou, Ms Sari Soderstrom (World Bank) and the collaborators in the applied research activities are especially thanked. One of us (VS) owes thanks to ACIL Australia for support given during the period 2007-2008.

## REFERENCES

Banks, T., Richard, C. Ping, L. Zhaoli. Y. 2003. Community-Based Grassland Management in Western China Rationale, Pilot Project Experience, and Policy Implications. *Mountain Research and Development* 23(2):132-140. 2003.

Bao, C and Fang, CL 2007. Water resources constraint force on urbanization in water deficient regions: A case study of the Hexi Corridor, arid area of NW China. *Ecological Economics 62(3-4)*:508-517.

Cao, S 2011 Impact of China's Large-Scale Ecological Restoration Program on the Environment and Society in Arid and Semiarid Areas of China: Achievements, Problems, Synthesis, and Applications. *Critical Reviews in Environmental Science and Technology*, 41(4): 317-335.

Feng, Z, Yang Y, Zhang Y, Zhang P, Li Y 2003. Grain-for-green policy and its impacts on grain supply in West China. *Land Use Policy* 22, 301-312.

Hua, LM and Michalk D.L 2010. Herders income and Expenditure: Perceptions and Expectations pp. 233-254 In: Victor Squires, Hua Limin, Zhang Degang and Li Guolin (eds) *Towards Sustainable Use of Rangelands in North-West China*. Springer, Dordrecht.

Hua, LM and Squires, VR 2015. Managing China's pastoral lands: current problems and future prospects. *Land Use Policy* 42: 129 137.

Hua, LM, Yang SW, Squires, VR and Wang GZ 2015. An Alternative Rangeland Management Approach in an Agro-Pastoral Area in Western China. *Rangeland Ecology and Management* 68:109–118 dx.DOI.org/ 10.1016/j.rama 2015.01.001.

Hua, LM and Zhang D. 2012. Engaging with land users: the first steps on a long road In: Victor Squires (ed) *Rangeland Stewardship in Central Asia: Balancing Improved Livelihoods, Biodiversity Conservation and Land Protection,* Springer, Dordrecht.

Jiang, Zehui (ed.) 2007. Integrated Ecosystem Management: *Proceedings of the International Workshop*. China Forestry Publishing House, Beijing.

Kemp, D and Michalk D 2011. Development of sustainable livestock systems on grasslands in north-western China. *Proceedings of an ACIAR workshop held at the combined International Grassland Congress and International Rangeland Conference,* Hohhot, Inner Mongolia Autonomous Region, China, 28 June 2008., pp. 12-24.

Lang, X, Wang C and Squires, V 2010. Protecting local breeds of livestock in NW China. pp.147-180 In: Victor Squires, Hua Limin, Zhang Degang and Li Guolin (eds) *Towards Sustainable Use of Rangelands in North-West China*. Springer, Dordrecht.

Li, WJ, Ali, SH, Zhang, Q. 2007. Property rights and grassland degradation: A study of the Xilingol Pasture, Inner Mongolia, China. Journal of Environmental Management. 85(2): 461-470.

Longworth, JW and Williamson. GJ. 1993. *China's Pastoral Regions: Sheep and Wool, Minority Nationalities, Rangeland Degradation and Sustainable Development.* Wallingford, UK: CAB International.

Lu Q, Wang S, Squires, V. and Yang Y. 2005. Desertification and dust storms in China: impacts, root causes and mitigation strategies. *China Forestry Science and Technology* 5(3):22-35.

Michalk, DL., Hua L., Kemp, D., Jones, R., Takahashi, T., Wu, J., Nan, Z, Xu, Z. and Han, G 2010. Redesigning Livestock Systems to Improve Household Income and Reduce

Stocking rates in China's western Grasslands pp.301-324 In: Victor Squires, Hua Limin, Zhang Degang and Li Guolin (eds) *Towards Sustainable Use of Rangelands in North-West China.* Springer, Dordrecht.

Sheehy, DP, Thorpe, J, Kirychuk, B. 2006. Rangeland, livestock and herders revisited in the northern pastoral region of China. *USDA Forest Service Proceedings* RMRS-P-39:62-68.

Squires, V. Lu X, LU Q Wang T and Yang, Y 2009 *Degradation and Recovery in China's Pastoral areas* CABI Wallingford 264p.

Squires, V. Hua L, Zhang, D and Li, G (eds) 2010. Towards Sustainable Use of Rangelands in North-West China. Springer, Dordrecht 353p.

Waldron, S, Brown, C and Zhao, Y. 2011. Developing the Right Institutional Environment to Deal with Grassland Degradation In: Kemp, D. and Michalk, D. (eds) *Sustainable Development of Livestock Systems on Grasslands in North-Western China. 21st International Grassland Congress and 8th International Rangelands Conference,* Hohhot, Inner Mongolia Autonomous Region, China, (pp.98-104). 28 June 2008.

Wang, L, and Wei H. 2003. The Progress and Effect Evaluation of Western Region Development in China. *Finance & Trade Economics,* No. 110 5-12. (in Chinese with English abstract).

Wang, MP, Zhao, CZ, Long R J, and Yang Y H. 2010. Rangeland governance in China: overview, impacts on Sunan County in Gansu Province and future options. *The Rangeland Journal* 32(2) 155–163.

World Bank 2003. Gansu and Xinjiang Pastoral Development Project. http: //www. worldbank.org/projects/P065035/gansu-xinjiang-pastoral-development- project.

Zhang, D., Ren, JZ, Hua, L. and Squires, V. 2010. Agro-pastoral integration in NW China: A New Paradigm pp.183-206 In: Victor Squires, Hua Limin, Zhang Degang and Li Guolin (eds) *Towards Sustainable Use of Rangelands in North-West China.* Springer Dordrecht.

Zhang, K and Wen, Z 2008. Review and challenges of policies of environmental protection and sustainable development in China. *Journal of Environmental Management,* 88(4):1249–1261.

Zhao, CZ, Hua, L. and Squires, V. 2010. Environmental Education: A Tool for Changing the Mind-set p.285-299 In: Victor Squires, Hua Limin, Zhang Degang and Li Guolin (eds) *Towards Sustainable Use of Rangelands in North-West China.* Springer, Dordrecht.

# Part 2. Rangeland Management in Practice

There are 4 chapters in this Part, all of which focus on experience in managing rangelands. Examples are drawn from widely divergent rangeland situations (Mediterranean Europe, Argentina, Mongolia and Tajikistan). Management objectives are different in each instance but the implementation of management plans and the measures taken to achieve this makes for interesting reading. Whether in the highly-regulated European agricultural scene or in the rural areas of Central Asia (Mongolia or Tajikistan) there are underlying principles to be observed and management interventions that can be 'make or break' such as manipulation of fire and grazing as in the Argentinean example.

In: Rangeland Ecology, Management ...
Editor: Victor R. Squires
ISBN: 978-1-63482-504-7


*Chapter 4*

# COMPARATIVE SUSTAINABILITY ASSESSMENT OF EXTENSIVE BEEF CATTLE FARMS IN A HIGH NATURE VALUE AGROFORESTRY SYSTEM

***A. J. Escribano[1], P. Gaspar[1,*], F. J. Mesías[2], M. Escribano[1] and F. Pulido[2]***

[1]Department of Animal Production and Food Science, University of Extremadura, Spain
[2]Department of Economics, University of Extremadura, Spain

## ABSTRACT

This chapter analyzes the sustainability of beef cattle systems of the Spanish Rangelands known as "Dehesas". These are traditional systems of the Iberian Peninsula where native herbaceous vegetation and evergreen species of Quercus provide the basis for extensive livestock farms. These systems are considered as outstanding High Nature Value (HNV) farming systems and the most extensive agroforestry systems in Europe according to CORINE Land Cover.

Beef farms in this area show low stocking rates and a small dependence on foodstuff purchases. However, certain changes have occurred in the last decades due to the Common Agricultural Policy (CAP). On the one hand, some farms have become more intensified, as a way to maximize the revenues from the CAP subsidies. On the other hand, many farms have turned to organic production, trying to take advantage both of new subsidies and of new market trends.

In this regard, the organic livestock farming model is gaining weight as an option for sustainable production since, according to various authors, these systems have advantages over conventional and intensive systems. In fact, organic production in the area has increased substantially in recent years due to several factors, such as the growing interest of the EU towards preserving sensitive ecosystems, the potential role of organic production in the development of rural areas and the growing consumers' demand for safer and higher quality foods produced under ethical and environmental standards.

In this study, we carried out a comparative assessment of the sustainability of different conventional and organic beef production systems located in dehesa rangelands.

* pgaspar@unex.es.

The systems analyzed were classified as follows. (i) non-organic farms (Conventional); (ii) organic farms that sold calves at weaning age as conventional ones (Organic 1); (iii) organic farms that fattened and sold their calves as organic (Organic 2). An adaptation of the MESMIS multicriteria framework was applied to calculate sustainability indices for each system. The results showed that the Organic 2 farms scored highest on most of the attributes of sustainability, as well as on the environmental and economic dimensions of sustainability. Thus, they were the most sustainable system (66.55%), followed by the Organic 1 (61.04%) and Conventional ones (56.89%). Organic 2 was the most sustainable system, but its implementation is complex due to both the high costs of organic inputs and the weak demand for organic beef.

The results also showed that all three types of systems need to improve in certain aspects that are crucial in the current and future context of the livestock sector. These aspects are: reducing the dependence on external feed, implementing more environmentally friendly farming practices, and farm diversification.

**Keywords:** sustainability, organic farming, livestock systems, beef cattle, Mediterranean, dehesa

# 1. INTRODUCTION

Pasture-based livestock farms are suffering from both intensification and abandonment due to several socioeconomic factors, such as the reduction of their profitability and competitiveness (Ripoll-Bosch et al., 2013). As a consequence, the sustainability of such farms and agro-ecosystems where they are located is jeopardized. Moreover, most of these systems are located in high value environmental ecosystems, such as the Spanish dehesas. (Mediterranean agroforestry systems)

Dehesa is an agroforestry system typical of the southwestern quadrant of the Iberian Peninsula. It is characterized by the presence of wooded pasture (the trees being predominantly species of oak), and by its main orientation to livestock extensive farming. Other uses on these systems also include crop farming (oats, barley and peas, mainly for re-use as animal feed), hunting and forestry (cork and firewood).

In the dehesa, the pastures, the trees, and the livestock complement each other and, when kept in equilibrium, conserve soil moisture, recharge the topsoil with nutrients and prevent the invasion of scrub (Schnabel, 1997; Montero et al., 1998). Indeed, the livestock plays a key role in the maintenance of the system as the use of appropriate cultural practices (such as a level of grazing suited to the system carrying forage production or forest regrowth) maintains the wooded layer, thus avoiding shrub invasion and increasing the system's efficiency (Coelho,1992).

These elements of grazing systems can remain in equilibrium for long periods of time, but imbalances can arise suddenly as the result of a critical change in one of the elements (Gaspar et al., 2009) affecting persistence of these systems.

In this context, it is thus necessary to design and implement livestock farming systems that take into account the abovementioned aspects with the objective of making them more sustainable. For this purpose, one necessary preliminary step is to unify the concept of sustainability in an agricultural context, as it can be different for different authors and stakeholders. According to Gómez-Limón and Sánchez-Fernández (2010), sustainable agricultural production is that which is economically viable (profitable), socially just (fair and equitable distribution of the wealth generated), and environmentally friendly (compatible with the maintenance of natural ecosystems). According to Rasul and Thapa (2004), it has to be low-input and regenerative, and has to make good use of a farm's internal resources through the incorporation of natural processes, knowledge, and improved production practices.

In this regard, the organic livestock farming model is gaining weight as an option for sustainable production since, according to various authors, these systems have advantages over conventional and intensive systems. Environmentally, it stands out that these systems apply practices that have less environmental impact on ecosystems and reduce their energy dependence to a greater extent than conventional (non-organic) systems (Tuomisto et al., 2012: Halberg 2012). Socioeconomically, this model can improve farm profitability by being able to demand a price premium (the difference in the price charged for the organic products with respect to the conventional ones).

However, the sustainability of production systems depends on many, often interrelated, factors which themselves vary from system to system and evolve over time (Ripoll-Bosch et al., 2012). One thus requires a holistic and dynamic framework that can cover the diversity of factors affecting sustainability, not just the legal framework defining organic production. This is especially important in areas which are fragile and with the complexity of a high number of trades-offs involving a wide range of stakeholders. The dehesa ecosystem is an example of such an area, since its management requires quite careful adjustment of the level of exploitation of the ecosystem. In fact, there is a fairly tight boundary between conservation and either over-exploitation or under-exploitation.

For these reasons, and given that there have as yet been no studies integrating participatory research and MESMIS techniques to compare the sustainability of organic and conventional beef cattle dehesa farms, the objective of the present work is to analyse the sustainability of these farms in a rural zone of southern Europe (Extremadura, SW Spain), in particular, those located in dehesas, using these methods.

## 2. Description of the Area of Study

The study area is the Autonomous Community of Extremadura in which dehesas are located (figure 1). This area is situated in the south-west quadrant of Spain, between lat 37º56'32''- 40º29'15" and long 4º38'52''- 7º32'35''. It is one of the main zones of dehesa in Spain. Indeed, approximately 50% of its useful agricultural area is considered dehesa, for a total of 2.2 million ha. The climate is continental Mediterranean. It presents annual mean temperatures ranging between 16-17° C. Summers are dry and hot (the mean temperature in July is over 26ºC, and the maximum is usually over 40ºC). Annual rainfalls are irregular during the year and also among years. The mean rainfall varies between 300 mm and 800 mm (Hernández, 1998).

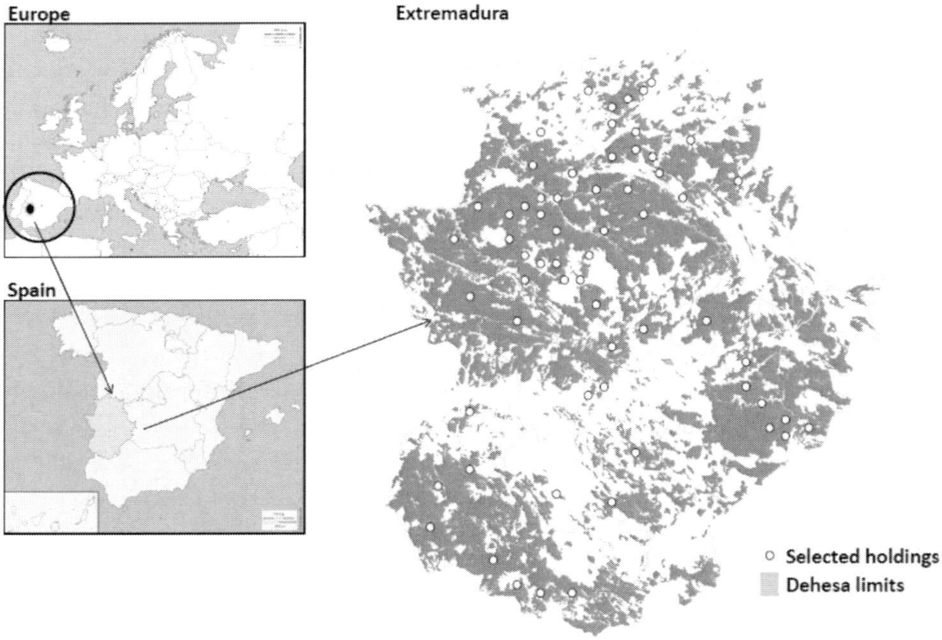


Figure 1. Extremadura region boundaries in Spain, dehesa limits in the region and location of selected farms for the study.

## 3. Overview of the Methodological Framework

The evaluation of the sustainability of the dehesa beef cattle farms in Extremadura Region was carried out in accordance with an adaptation of the MESMIS framework (Management System Evaluation Framework incorporating Sustainability Indicators) proposed by Masera et al. (1999). The adaptation we made for the present study allowed the characteristics of dehesa farms, the organic production model established by Council Regulation (EC) Nº 834/2007, the current challenges facing the livestock sector (increasing input prices and competitiveness) and the trends of the future CAP (quality products, and environmental and social services) to be integrated into the evaluation of the systems' sustainability. MESMIS is readily adaptable to whichever agro-ecosystems are under study. It permits sustainability indicators to be derived, measured, and monitored as part of a systemic, participatory, interdisciplinary, and flexible process of evaluation (López-Ridaura et al., 2002). The MESMIS framework has demonstrated its power to identify indicators for the evaluation of different agro-ecosystems and livestock systems in particular nearly 15 years (Astier et al., 2012; Ripoll-Bosch et al., 2012). However, the present work seems to be the first to apply it to comparing organic and conventional beef cattle farms.

The methodological framework of the research is presented in figure 2 and includes the following four steps: (i) selection of beef cattle systems, (ii) selection of sustainability indicators, (iii) data collection and database creation and (vi) sustainability assessment.

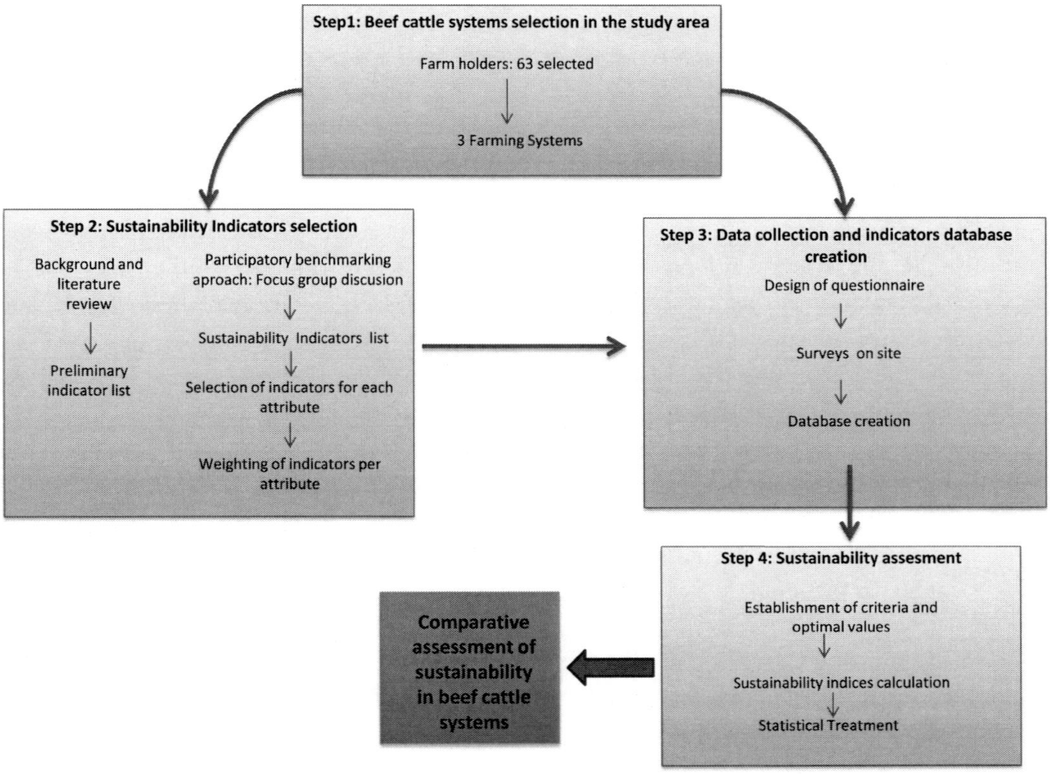


Figure 2. Methodological flowchart for studying the sustainability dehesa beef cattle systems.

## 3.1. Beef Cattle Systems Selection in the Study Area (Step 1)

This study was carried out within the research project 'Study of the viability of different models of organic beef production: Influence on meat quality and economic analysis' (code INIA-RTA2009-00122-C03-03). As one of the objectives of the project was the comparison of conventional and organic beef cattle farms in Extremadura (SW Spain), similar number of conventional and organic farms were selected, in order to make such comparison possible. The farms analyzed in this study are all located in the dehesa ecosystem. In order to select the farms, preliminary lists were obtained from the certification body of organic production and from the main cooperatives of the region. Four selection criteria were taken into account to select the final holders included in the study:

- Beef cattle must be predominant over other livestock species.
- Farms' dimension must be over 25 adult cows, in order to avoid non-professional farms.
- Selected farms should be representative of the different types of dehesa: woodland density, orography, soil conditions and pastures production. This requisite has been used in several studies addressing livestock production systems (Milán et al., 2006; Gaspar et al., 2007; Gaspar et al., 2008; Gaspar et al., 2009a; Gaspar et al. 2009b).

- All organic farms must have overcome the conversion period to the organic production system[1].

Finally, 63 farms were selected ( 30 conventional farms and 33 organic farms), which is in line with other studies analyzing livestock production systems (Serrano et al., 2004; Gaspar et al., 2007; Gaspar et al., 2009a; Gaspar et al. 2009b). As some of the organic farms selected are behaving almost conventional farms (they are certified as organic but their products are sold in the conventional market to avoid the constraints of organic markets), three beef cattle systems were identified:

- System 1 ('Conventional'): It includes the conventional farms n= 30.
- System 2 ('Organic 1'): comprised 22 farms that were certified as organic, whose purpose was the production and sale of non-organic calves at the age of weaning (4.5-6 months of age and 160-220 kg live weight) to be fattened on other farms (under the conventional system). Although the farms in this system are certified as organic, their output is sold as conventional, since there is no demand for organic weaned calves.
- System 3 ('Organic 2'): This system includes 11 certified organic farms, that complete the fattening of all their calves in accordance with the organic model, and sell all of them as organic ones.

## 3.2. Sustainability Indicators Selection (Step 2)

### *3.2.1. Background and Literature Review*

Firstly, a preliminary list of sustainability indicators were designed on the basis of the scientific literature (Nahed et al., 2006; Gaspar et al., 2007; Gaspar et al., 2008; Gaspar et al., 2009a; Gaspar et al., 2009b; Blanco-Penedo et al., 2012; Gómez-Limón and Sánchez-Fernández, 2010; Lebacq et al., 2013; Ripoll-Bosch et al., 2012; Ripoll-Bosch et al., 2013), on EAA (2001, Regulation EC No 138/2004 and subsequent amendments), and on the Council Regulation (EC) No. 834/2007.

### *3.2.2. Participatory Benchmarking Approach: Focus Group Discussion*

Secondly, a participatory research strategy based on a focus group was applied. The focus group comprised 4 farm managers, 3 technicians (veterinarians and agronomists), and 5 academic experts. It allowed integrating local and scientific knowledge, taking into account stakeholders interests and values, following a collaborative procedure among researchers, advisers, and farmers.

In a first meeting the focus group selected the factors considered important for the sustainability of beef cattle farms, of dehesa agro ecosystem, and of the rural populations. The previous list of indicators was then presented to the participants who evaluated the appropriateness, relevance and measurability of the indicators selected, according to the

[1] According to the Council Regulation 834/2007, this process last 3 years for beef cattle. During such period, farms comply with more than the 80% of the legal requirements needed to be organic.

recommendations of Lebacq et al. (2013) and Dale et al. (2013). Finally, a definitive list of indicators was set on the basis of three criteria: (i) they were currently used in similar sustainability studies; (ii) they were coherent with the environmental social, economic, and rural orientation of the study; and (iii) they addressed present and future problems of the sustainability of rangeland beef cattle production systems, together with dehesa and organic production.

In order to increase the practical applicability of the indicators selected, they were grouped into 'attributes' selected from the scientific literature. Once indicators were grouped within attributes, relative weights were assigned to each indicator. Thus, the sum of the relative weights of each indicator of the same attribute will be equal to 100%.

In the following paragraphs, we shall describe the attributes on whose basis the sustainability was evaluated.

*Adaptability and Flexibility:* This is the system's capacity to find new levels of equilibrium, i.e., to continue being productive or profitable when faced with long-term changes in the environment (e.g., new economic or biophysical conditions). This attribute also includes the capacity to actively seek new levels or strategies of production when facing critical points to their adaptability.

*Self-reliance*: This is the system's capacity to regulate and control its interactions with the exterior.

*Equity:* This is the system's capacity to distribute wealth, costs, and resources fairly, both within and between generations. It also relates to the quality of life. Its indicators are measures related to workforce, as well as access to services and information, and the degree of social interaction.

*Stability and Resilience:* This refers to the system's capacity to return to equilibrium or productivity levels similar to the initial level after serious disturbance, and to maintain the profits provided by the system at a non-declining level over time, under normal or average conditions.

*Productivity:* This is the agro-ecosystem's capacity to provide the required level of goods and services. It represents the value of the attribute in a given time period. It is a key indicator of the economic dimension of sustainability.

The selected indicators, their units, the attribute in which they were grouped and their weights are presented in Table 1.

## 3.3. Data Collection and Indicators Database Creation (Step 3)

On the basis of the indicators selected, a questionnaire was designed. The questionnaire included items corresponding to both qualitative and quantitative aspects that were necessary in order to calculate indicators included in the list. Information was collected by direct surveys to the farm manager and by personal observation directly on the farm. Afterwards a database was created with the purpose of perform the analyses.

**Table 1. Selected attributes and indicators: definitions, units, and weights**

| Attributes and indicators | | Units* | Criteria** | Weight |
|---|---|---|---|---|
| Adaptability and Flexibility | | | | |
| | Number of agricultural activities/products | Qualitative scale | Max. | 20.03 |
| | Wooded UAA per total UAA | % | Max. P90 | 19.72 |
| | Dependence on livestock revenue | % | Min. P10 | 19.00 |
| | Farmer's age | Years | Rec. | 15.03 |
| | Level of studies | Qualitative scale | Max. | 12.70 |
| | Off farm income/total income | % | Max. | 7.98 |
| | Cows per bull | Cows | Q1 | 5.53 |
| Self-reliance | | | | |
| | Owned UAA per total UAA | % | Max. | 18.30 |
| | Dependence on subsidies | Subsidies per total income. % | Min. | 18.10 |
| | External dependence on animal feed | €/ha | Q1 | 17.30 |
| | Family AWU/Total AWU per 100 ha UAA | AWU100 $ha^{-1}$ | Q3 | 13.30 |
| | Cultivated UAA per total UAA | % | Q3 | 10.90 |
| | Quality of tracks | Qualitative scale | Max. | 8.90 |
| | Veterinary and medicine expenses | €$ha^{-1}$ | Q1 | 7.10 |
| | Other intermediate consumption | €$ha^{-1}$ | Q1 | 6.10 |
| Equity | | | | |
| | Workforce stability | AWU100$ha^{-1}$ | Q3 | 24.60 |
| | Total AWU per 100 ha UAA | AWU100 $ha^{-1}$ | Q3 | 17.90 |
| | Number of external workers | AWU100 $ha^{-1}$ | Q3 | 14.70 |
| | Satisfaction level | Qualitative scale | Max. | 13.40 |
| | Distance to social services | Km. | Q1 | 11.02 |
| | Cattlemen associations membership | Qualitative scale | Max. | 10.18 |
| | Farmer's gender | Dichotomous | Yes. | 4.30 |
| | Distance to the slaughterhouse | Km. | Q2 | 3.90 |
| Stability and resilience | | | | |
| | Total stocking rate | $LUha^{-1}$ | Rec. | 21.50 |
| | Farm continuity/future plans | Qualitative scale | Max. | 19.30 |
| | Soil and crop management | Qualitative scale | Max. | 12.67 |
| | Use of pesticides and/or mineral fertilizers | Dichotomous | No | 11.23 |
| | Degree of integration | Dichotomous | Yes | 10.30 |
| | Use of preventive antiparasitics and/or antibiotics | Dichotomous | No | 8.10 |
| | Dung management | Qualitative scale | Max. | 7.90 |
| | Carbon sequestration | C $tonsha^{-1}$ | Q3 | 4.90 |
| | Percentage of autochthonous ruminants | % | Q3 | 4.10 |
| Productivity | | | | |
| | Livestock productivity | Total sales/total LU. €$LU^{-1}$ | Q3 | 15.24 |
| | Labour productivity | Total sales/total AWU. €$AWU^{-1}$ | Q3 | 15.07 |
| | Profitability rate | % | Q3 | 12.90 |
| | Land productivity | €$ha^{-1}$ | Q3 | 12.89 |
| | Cow productivity | Calves | Rec. | 12.30 |
| | Net value added | €$ha^{-1}$ | Q3 | 11.20 |
| | Sales of livestock | €$ha^{-1}$ | Q3 | 9.00 |
| | Other sales | €$ha^{-1}$ | Q3 | 6.90 |
| | Mortality rate | % | Min. | 4.50 |

* Dichotomous: yes/no; UAA: Utilized Agricultural Area; AWU: Annual work Units; LU: Livestock Units;

** Max.: maximum value; Min.: minimum value; Q3: third quartile; Q2: second quartile; Q1: first quartile; P90: 90th percentile.

## 3.4. Sustainability Assessment (Step 4)

### *3.4.1. Establishment of Criteria for Optimal Values*

Once the data were collected the optimal or reference values were established. These are desirable values for each indicator attainable under the circumstances on an ideal system (Nahed et al., 2006). The establishment of the optimal values is one of the most critical points in the assessment of the farms' sustainability (Masera et al., 1999), since the results, conclusions and measures implemented at the farm level will depend on such values. For establishing the optimal values, different principles (criteria) must be followed. Thus, optimal values can be selected by taking into account the maximum value of the sample, certain quartile, percentile or a value recommended in the scientific literature.

The optimal values were set in a new meeting of the focus group. The participants were given a table of all the indicators and their basic statistical descriptors (maximum, minimum, mean, and percentiles). They were also provided with information regarding the values used in other research papers (Nahed et al., 2006; Gaspar et al., 2009a; Gaspar et al., 2009b; Ripoll-Bosch et al., 2012)

The attendees were asked to choose one of the values given as optimal for each indicator, and only in extreme cases to give another value. The results were summarized and presented to the focus group again, to try to reach greater consensus. The criteria used for each indicator is also presented in Table 1.The optimal values are shown in Tables 2 and 3 in order to present them along with the results of the sustainability indicators.

### *3.4.2. Sustainability Indices Calculation*

The values of the indicators were transformed into sustainability indices. These indices are expressed as percentages, ranging from 1 to 100. Thus, the selected farms obtained scores that allowed a better understanding of their relative sustainability, since the closer they are to 100, the better the system in terms of sustainability. Such conversion was made on the basis of the AMOEBA procedure (Ten Brink et al., 1991), by using the optimal value of each indicator as follows:

- For the indicators whose optimal values were chosen to be the maximum, the index was calculated by means of the equation (1):

$$\text{Sustainability index} = (\text{indicator value} / \text{optimal value}) \times 100 \qquad (1)$$

- For the indicators whose optimal values were to be the minimum, the index was calculated by means of the equation (2):

$$\text{Sustainability index} = (\text{optimal value} / \text{indicator value}) \times 100 \qquad (2)$$

For the indicators whose values were lesser than the optimal value and whose optimal values were chosen to be some percentile, the mean, or a recommended value; formula 1 was applied. Conversely, formula 2 was used when the value of the indicator was greater than the optimal value.

### *3.4.3. Statistical Treatment of Results*

To detect differences between farms' systems sustainability indices (scores), a single factor one-way analysis of variance (ANOVA) was applied. These analyses were carried out with the 2011 Statistical Package for Social Systems, version 20.0.

# 4. RESULTS

The results for the sustainability indicators for the three farm systems are presented in Tables 2-4. Table 2 shows the mean values and standard deviation of the quantitative indicators. Table 3 shows the results for the qualitative indicators. Table 4 presents the calculated sustainability indices (scores).

**Table 2. Mean values and standard deviation of the quantitative sustainability indicators for the three farm systems**

| Attributes and indicators | | Conventional | Organic 1 | Organic 2 | Optimal | Sample (±S.D.) |
|---|---|---|---|---|---|---|
| Adaptability and Flexibility | | | | | | |
| | Wooded UAA per total UAA | 45.56 | 46.00 | 75.80 | 98.00 | 50.98 (±42.40) |
| | Dependence on livestock revenue | 60.50 | 53.14 | 51.27 | 37.95 | 56.32 (±17.98) |
| | Farmer's age | 45.94 | 46.59 | 50.73 | 30.00 | 47.00 (±9.98) |
| | Off farm income/total income | 17.92 | 8.41 | 29.55 | 100.00 | 16.63 (±28.44) |
| | Cows per bull | 30.59 | 30.67 | 28.29 | 20.88 | 30.22 (±13.40) |
| Self-reliance | | | | | | |
| | Owned UAA per total UAA | 64.43 | 54.45 | 55.27 | 100.00 | 59.35 (±44.40) |
| | Dependence on subsidies | 11.13 | 12.73 | 12.27 | 0.00 | 11.89 (±21.73) |
| | External dependence on animal feedstuff | 109.69 | 17.55 | 96.63 | 6.71 | 75.24 (±112.18) |
| | Family AWU/Total AWU per 100 ha UAA | 83.91 | 79.90 | 48.80 | 100.00 | 76.38 (±31.70) |
| | Cultivated UAA per total UAA | 11.13 | 12.73 | 12.27 | 13.38 | 11.89 (±21.73) |
| | Veterinary and medicine expenses | 17.87 | 4.51 | 4.84 | 1.41 | 10.93 (±21.16) |
| | Other intermediate consumption[a] | 145.43 | 117.64 | 122.57 | 61.13 | 131.73 (±162.27) |
| Equity | | | | | | |
| | Workforce stability[b] | 0.16 | 0.77 | 0.55 | 0.19 | 0.20 (±0.37) |
| | Total AWU per 100 ha UAA | 1.12 | 0.85 | 0.67 | 1.05 | 0.95 (±1.04) |
| | Number of external workers | 0.39 | 0.25 | 1.23 | 0.95 | 0.49 (±0.74) |
| | Distance to social services[c] | 22.27 | 33.39 | 41.08 | 13.60 | 29.44 (±23.78) |
| | Distance to the slaughterhouse | 19.20 | 68.30 | 95.36 | 35.40 | 49.65 (±44.26) |
| Stability and resilience | | | | | | |
| | Total stocking rate | 0.73 | 0.50 | 0.44 | 0.33 | 0.60 (±0.64) |
| | Carbon sequestration[d] | 0.05 | 0.05 | 0.08 | 0.07 | 0.05 (±0.03) |
| | Percentage of autochthonous ruminants | 23.73 | 34.28 | 66.63 | 86.02 | 34.90 (±42.67) |

| Attributes and indicators | | Conventional | Organic 1 | Organic 2 | Optimal | Sample (±S.D.) |
|---|---|---|---|---|---|---|
| Productivity | | | | | | |
| | Livestock productivity | 429.32 | 350.43 | 556.81 | 620.92 | 424.03(±224.24) |
| | Labour productivity | 37,253 | 21,958 | 37,805 | 54,673 | 32,008 (±20,798) |
| | Profitability rate[e] | 4.43 | 4.20 | 4.35 | 5.76 | 4.34 (±2.69) |
| | Land productivity[f] | 642.40 | 470.25 | 586.65 | 576.67 | 572.55 (±410.82) |
| | Cow productivity | 0.81 | 0.70 | 0.65 | 1.00 | 0.75 (±0.21) |
| | Net value added[g] | 383.00 | 333.40 | 375.22 | 423.07 | 364.32 (±245.07) |
| | Sales of livestock | 292.96 | 152.84 | 215.92 | 244.28 | 230.58 (±223.23) |
| | Other sales | 1.49 | 11.34 | 9.75 | 0.09 | 6.37 (±23.05) |
| | Mortality rate | 3.12 | 4.34 | 10.21 | 0.00 | 4.79 (±9.49) |

S.D.: Standard deviation.

[a] It is the sum of workforce salaries + amortization costs + land leasing costs.

[b] Workforce stability: [((Fixed AWUs)/ (Fixed AWUs + Temporary AWUs)]) × 100.

[c] Distance to the nearest municipality with at least 10,000 inhabitants.

[d] The methodology followed by Murillo et al. (2009) was used. Tons of carbon stored = estimated farm's total forest biomass per ha × carbon stored by photosynthesis. Estimated farm's total forest biomass was based calculated by integrating the characteristics of the farms' woody area (species, density and size) and the data provided by the 3rd National Forest Inventory (MAGRAMA, 2007). Carbon stored by photosynthesis = 0.5. Estimated farm's total forest biomass / ha = (estimation abovementioned / UAA) ×* BEF ×* WD. BEF: biomass expansion factor = 1.69. WD: wood density of Olm oaks and Quercus sp.= 0.5. It is the coefficient use to convert the biomass to dry matter.

[e] Ratio between net surplus and the average capital assets, estimated from the value of total fixed capital and the value of capital.

[f] Gross output/ha UAA. All the products of the agricultural activities. All agricultural output is recorded except that produced by units that produce solely for their own consumption.

[g] Measures the value created by all the agricultural output after the consumption of fixed capital. That output is valued at basic prices and intermediate consumption is valued at purchaser prices. Is calculated as follows: (Gross output – Intermediate consumption – Amortization) + (those subsidies not related to livestock farming).

**Table 3. Qualitative sustainability indicators for the three farm systems: categories of each indicator and percentage of farms**

| | Attributes and indicators | Conventional | Organic 1 | Organic 2 | Optimal |
|---|---|---|---|---|---|
| Adaptability and Flexibility | Number of agricultural activities/products | | | | |
| | *1: One activity* | 86.70 | 59.10 | 54.50 | |
| | *2: Two activities* | 23.10 | 53.80 | 23.10 | |
| | *3: Three activities* | 0.00 | 9.10 | 0.00 | 5 |
| | *4: Four activities* | 3.20 | 0.00 | 9.10 | |
| | *5: Five activities* | 0.00 | 0.00 | 9.10 | |
| | Level of studies | | | | |
| | *1: No studies or basic education* | 30.00 | 40.90 | 27.30 | |
| | *2: Secondary and/or vocational education and training* | 30.00 | 36.40 | 36.40 | 3 |
| | *3: University degree* | 40.00 | 22.70 | 36.40 | |

**Table 3. (Continued)**

| | Attributes and indicators | Conventional | Organic 1 | Organic 2 | Optimal |
|---|---|---|---|---|---|
| Self-rel. | Quality of tracks | | | | |
| | *1: Good* | 63.30 | 72.70 | 72.70 | |
| | *2: Need improvement* | 26.70 | 27.30 | 27.30 | 1 |
| | *3: Bad quality* | 10.00 | 0.00 | 0.00 | |
| Equity | Satisfaction level | | | | |
| | *1: Unsatisfied* | 46.70 | 40.90 | 45.50 | |
| | *2: Intermediate level of satisfaction* | 46.70 | 50.00 | 54.50 | 3 |
| | *3: Satisfied* | 6.70 | 9.10 | 0.00 | |
| | Cattlemen association membership | | | | |
| | *1: Belong to 0 associations* | 6.00 | 8.11 | 0.00 | |
| | *2: Belong to 1 association* | 20.00 | 20.00 | 0.00 | 3 |
| | *3: Belong to 2 or more associations* | 74.00 | 64.86 | 100.00 | |
| | Farmer's gender[a] | | | | |
| | *1: Female* | 3.30 | 18.20 | 9.10 | 1 |
| | *2: Male* | 96.70 | 81.80 | 90.90 | |
| Stability and resilience | Farm continuity/future plans | | | | |
| | *1: Abandon* | 6.70 | 4.50 | 0.00 | |
| | *2: Herd reduction or change of species* | 3.30 | 0.00 | 9.10 | |
| | *3: Conversion to the conventional system or conversion to the organic* | 60.00 | 81.80 | 72.70 | 3 and 4 |
| | *4: Increase the herd size* | 30.00 | 13.60 | 18.20 | |
| | Soil and crop management[b] | | | | |
| | *0: None measure/ is carried out* | 16.70 | 13.60 | 0.00 | |
| | *1: One measure is implemented* | 16.70 | 31.80 | 36.40 | |
| | *2: Two measures* | 46.70 | 27.30 | 45.50 | 5 |
| | *3: Three measures* | 16.70 | 18.20 | 0.00 | |
| | *4: Four measures* | 0.00 | 9.10 | 9.10 | |
| | *5: Five measures* | 3.30 | 0.00 | 9.10 | |
| | Use of pesticides and/or mineral fertilizers | | | | |
| | *0: No* | 63.30 | 100.00 | 100.00 | 0 |
| | *1: Yes* | 10.00 | 0.00 | 0.00 | |
| | Degree of integration | | | | |
| | *0: No integration of various livestock species, crops and trees* | 93.30 | 86.40 | 63.30 | 1 |
| | *1: Integrate these elements* | 6.70 | 13.60 | 36.40 | |
| | Use of preventive antiparasitics and/or antibiotics | | | | |
| | *0: They do not use them* | 3.30 | 100.00 | 100.00 | 0 |
| | *1: Yes, they use them* | 96.70 | 0.00 | 0.00 | |
| | Dung management | | | | |
| | *1: No dung due to extensiveness* | 30.00 | 36.40 | 0.00 | |
| | *2: Spreading of immature dung* | 56.70 | 45.50 | 45.50 | 4 |
| | *3: Dung heaping (immature)* | 13.30 | 4.50 | 45.50 | |
| | *4: Compost* | 0.00 | 13.60 | 9.10 | |

[a]From the sustainable rural development points of view, the participation of women in the management and ownership of farms, play a double role. On the one hand, it is more equitable, since there is no exclusion of women from the rural world. On the other hand, the fixation of women in rural areas can be related to the establishment of families and more population in such areas.

[b] Number of measures/agricultural management practices implemented to reduce soil erosion and to improve soil quality. These include: cover crops, mulching, intercropping, crop rotation, plot rotation, fallow, and use of compost.

In the following paragraphs, we shall describe the main findings of the farm systems. Table 4 shows the calculated sustainability indices (scores) of each farm system and of the sample.

## 4.1. Adaptability and Flexibility

As can be seen in Table 4, significant differences were found among the farm systems ($p<0.01$). These differences were mainly due to the number of agricultural activities/products and the wooded UAA per total UAA, as both indicators are interlinked. More wooded area allows a higher degree of diversification of the holding (measured as the number of agricultural activities/products).The degree of business diversification and the wooded area are key elements for the dehesa farms. In this sense, the Organic 2 systems showed the greatest degree of diversification (54.55%), and of wooded area (77.34%). As a result, the set of indicators studied showed the Organic 2 farms to be more adaptable and flexible than the Conventional and Organic 1 farms.

## 4.2. Self-Reliance

Organic 2 farms showed the lowest percentage of family AWU / total AWU per 100 ha UAA, and the worst scores (48.80%). This may be because their specialization and their higher degree of diversification force the farm-holder to hire outside labour, thus reducing the farm's self-reliance. Additionally, reducing outlays on feed is especially important since it usually represents the greatest cost of production. This is especially important in organic farms, since the prices of organically grown feedstuffs are much higher than conventional ones. In this sense, the Organic 1 farms scored best (68.27%) in terms of their relative lack of dependence on external feed. This is because they did not fatten their calves, but they sold them on at weaning, so that they hardly consumed any concentrate. On these farms, only the breeding cows and bulls required externally supplied feed at times when the available food resources generated by the farm (pasture, hay, and crops produced on the farm itself) do not cover the animals' nutritional needs. The Conventional farms scored next best (33.46%). This is because not all the farms in this farm system fattened their calves and the conventional feedstuff price is lower. Organic 2 farms had the worst scores because they all fattened their calves organically. In summary, with respect to the self-reliance attribute the Organic 1 system showed to be the best positioned, followed by Organic 2, and Conventional the worst positioned (70.14%, 61.05% and 57.99%, respectively).

In the case of the Organic 1 farms, the differences were strongly influenced by their not being full-cycle production systems, meaning that they had reduced feed purchase and labour costs.

**Table 4. Mean values, standard deviation and significance of the sustainability indices (scores) for the three farm systems**

| Attributes and indicators | Convent. | Organic 1 | Organic 2 | Sample (±S.D.) | P-value |
|---|---|---|---|---|---|
| Adaptability and Flexibility | 54.14 [a] | 56.42 [a,b] | 65.34 [b] | 56.89 (±12.44) | * |
| Number of agricultural activities/products | 38.89 [a] | 50.00 [a,b] | 54.55 [b] | 45.50 (±21.00) | * |
| Wooded UAA per total UAA | 46.48 | 46.92 | 77.34 | 52.02 (±43.27) | NS |
| Dependence on livestock revenue | 66.39 | 78.45 | 76.15 | 72.31 (±18.91) | NS |
| Farmer's age | 68.38 | 65.95 | 61.64 | 66.35 (±13.96) | NS |
| Level of studies | 70.00 | 60.61 | 69.70 | 66.67 (±27.44) | NS |
| Off farm income/total income | 17.92 | 8.41 | 29.55 | 16.63 (±28.44) | NS |
| Cows per bull | 71.75 | 71.62 | 76.24 | 72.49 (±20.15) | NS |
| Self-reliance | 61.05 [a] | 70.14 [b] | 57.99 [a] | 63.69 (±13.01) | * |
| Owned UAA per total UAA | 64.45 | 54.42 | 55.35 | 59.36 (±44.42) | NS |
| Dependence on subsidies | 88.84 | 87.38 | 87.95 | 88.17 (±21.71) | NS |
| External dependence on animal feedstuff | 33.46 [a] | 68.27 [b] | 26.06 [a] | 44.33 (±44.64) | ** |
| Family AWU/Total AWU per 100 ha UAA | 83.91 [a] | 79.90 [a] | 48.80 [b] | 76.38 (±31.70) | ** |
| Cultivated UAA per total UAA | 27.92 | 49.37 | 50.96 | 39.43 (±43.82) | NS |
| Quality of tracks | 76.67 | 86.36 | 86.36 | 81.75 (±28.82) | NS |
| Veterinary and medicine expenses | 30.80 [a] | 65.98 [b] | 63.09 [b] | 48.72 (±37.52) | ** |
| Other intermediate consumption | 68.39 | 68.45 | 52.91 | 65.71 (±28.25) | NS |
| Equity | 57.90 | 55.18 | 63.44 | 55.53 (±14.72) | NS |
| Workforce stability | 10.00 [a] | 4.55 [a] | 45.45 [b] | 14.29 (±35.27) | ** |
| Total AWU per 100 ha UAA | 77.98 [a] | 60.48 [b] | 62.88 [b] | 62.44 (±29.44) | * |
| Number of external workers | 23.99 [a] | 22.34 [a] | 49.78 [b] | 27.91 (±36.35) | * |
| Satisfaction level | 52.87 | 55.59 | 51.00 | 53.49 (±20.18) | NS |
| Distance to social services | 77.73 [a] | 66.61 [b] | 58.92 [b] | 70.56 (±23.78) | * |
| Cattlemen associations membership | 85.00 | 79.55 | 100.00 | 85.71 (±27.57) | NS |
| Farmer's gender | 51.67 | 59.09 | 54.55 | 54.76 (±14.80) | NS |
| Distance to the slaughterhouse | 96.77 [a] | 61.30 [b] | 44.93 [b] | 75.33 (±29.13) | *** |
| Stability and resilience | 43.46 [a] | 61.85 [b] | 69.31 [b] | 54.39 (±16.49) | *** |
| Total stocking rate | 61.47 | 69.71 | 69.46 | 65.74 (±22.71) | NS |
| Farm continuity/future plans | 85.67 | 87.27 | 88.18 | 86.67 (±21.33) | NS |
| Soil and crop management | 35.33 | 35.45 | 41.82 | 36.51 (±23.43) | NS |
| Use of pesticides and/or mineral fertilizers | 55.00 [a] | 100.00 [b] | 100.00 [b] | 78.57 (±39.87) | *** |
| Degree of integration | 6.67 [a] | 13.64 [ab] | 36.36 [b] | 14.29 (±35.27) | * |
| Use of preventive antiparasitics and/or antibiotics | 3.33 [a] | 100.00 [b] | 100.00 [b] | 53.97 (±50.24) | *** |
| Dung management | 28.33 | 36.36 | 54.55 | 35.71 (±40.59) | NS |
| Carbon sequestration | 63.74 [a] | 65.30 [a] | 97.31 [b] | 70.15 (±41.70) | * |
| Percentage of autochthonous ruminants | 25.48 [a] | 36.16 [ab] | 70.50 [b] | 37.07 (±44.38) | * |
| Productivity | 67.93 [a,b] | 61.62 [a] | 76.69 [b] | 67.26 (±16.84) | * |
| Livestock productivity | 74.40 [a,b] | 64.60 [a] | 89.93 [b] | 73.69 (±24.97) | * |
| Labour productivity | 67.46 [a] | 46.42 [b] | 74.26 [a] | 61.30 (±31.25) | * |
| Profitability rate | 62.65 | 65.39 | 72.56 | 65.34 (±31.13) | NS |
| Land productivity | 75.39 [a,b] | 71.31 [a] | 93.24 [b] | 77.08 (±23.11) | * |
| Cow productivity | 80.92 | 70.44 | 65.12 | 74.50 (±21.03) | NS |
| Net value added | 70.46 | 67.45 | 83.46 | 71.68 (±25.96) | NS |
| Sales of livestock | 68.83 [a,b] | 56.25 [a] | 86.23 [b] | 67.48 (±29.22) | * |
| Other sales | 3.33 | 22.73 | 18.18 | 12.70 (±33.56) | NS |
| Mortality rate | 96.88 | 95.66 | 89.79 | 95.21 (±9.49) | NS |
| Overall sustainability | 56.89 [a] | 61.04 [b] | 66.55 [c] | 60.03 (±6.87) | *** |

[a, b, c] Values within a row with different superscripts differ significantly at P<0.05.

Pvalue; NS: P>0.05; *=P<0.05; **=P<0.01; ***=P<0.001.

### 4.3. Equity

In relation to the attribute of equity, the first three indicators are related to labor, since it is considered that the more workers employed in the farms (preferably non family worker), the more equitable the systems are. Conventional farms are those that employ more people, although organic farms (both systems) hire more external workers, making therefore a higher contribution to the distribution of the profits obtained from the farms.

Regarding the access to social services (health, leisure, etc.), the distance between the farm and the nearest municipality with at least 10,000 inhabitants was been used as an estimator. Social services are available only in large municipalities in the area. Thus, the longer the distance between the farms and the municipalities, the more time is needed to access the public services, thus reducing the farmers' quality of life. In this sense, Conventional farms scored the highest (77.73%). Similarly, differences ($p<0.001$) between the distance to the slaughterhouse between Conventional and Organic farms were found. Conventional farms were located closer to these facilities, due to the fact that there are few organically certified slaughterhouses in the region (only 2).

Overall, no farms system presented significantly better scores for the Equity attribute.

### 4.4. Stability and Resilience

As can be observed in Table 4, remarkable differences ($p<0.001$) were found among Conventional and Organic farms. As the majority of indices grouped in this attribute are related to the agro-ecosystem management, these differences highlight that Organic farms contribute to environmental stability to a greater extent than conventional farms do. Specifically, significant differences ($P<0.001$) were found among Conventional and Organic farms (Organic 1 and Organic 2) for the use of pesticides and/or mineral fertilizers (55.00% vs. 100.00%), and for the use of preventive anti-parasitics and/or antibiotics (3.33% and 100.00%). This result is due to the fact that the EU legislation in force on organic production (Council Regulation EC Nº 834/2007 and subsequent amendments) prohibits the use of these products.

However, for the degree of integration (measured as the integration of different livestock species, crops, and trees), the differences were found among the Organic 2 system and the rest of systems ($p<0.05$). The higher degree of integration can be explained by the fact that Organic 2 farms are mainly located in the centre and South of Extremadura, were soils are of higher quality, and the density of trees and their size is also higher. This allows integrating crops and other livestock species (such as the Iberian pigs) more easily.

### 4.5. Productivity

Regarding the indicators with more weight on the attribute of productivity, (livestock and labour productivity) Organic 2 farms obtained the highest scores because they fattened all their livestock (so that there was greater retention of value added) and because of the higher price at which organic calves are sold (25% above the price of the fattened calves of the

Conventional system.) Accordingly the Organic 1 system positioned last (61.62%) comes from the fact that they do not fatten their calves as they mainly try to maximize subsidies.

Considering the overall sustainability of the systems analysed, the Organic 2 farms are the most sustainable, followed by Organic 1, and Conventional.

## 5. DISCUSSION

The degree of business diversification is a key element in rangeland farms in general, and in dehesa farms in particular (Gaspar et al. 2009b), because their profitability is low and their economic stability is strongly dependent on the farm's feeding resources. In this sense, farms with a greater variety of outputs (livestock, olives, firewood, cork, etc.)such as Organic 2 farms have a greater capacity to adapt to changes such as the collapse of prices in the beef cattle sector. This degree of diversification is reinforced by the higher percentage of wooded area that this farm system showed.

The availability of wooded area is a key element for dehesa farms. Economically, it allows them to increase their feed resources and their diversification potential, increasing the economic performance and reducing the economic risk. Environmentally, the presence of wooded areas reduces erosion, leaching, and groundwater pollution, increases biodiversity, and improves nutrient recycling (Smith et al. 2011). In this sense, it would be advisable for dehesa farms to increase their wooded areas and, in some cases, the density of trees, as this would redound in environmental and economic benefits.

With regard to the self-reliance it is interesting to note that with the high prices of feedstuff and the projected constraints that pasture-based livestock systems of Mediterranean basin can suffer due to climate changes (Nardone et al. 2010, Segnalini et al. 2013), self-reliance becomes a key aspect for dehesa systems, which must improve their performance with respect to it. In this sense, Organic 1 farms had the highest scores due to their reduced use of concentrates, since they base their production system on farms' natural resources. Despite this appearing to be advantageous, it may undermine other sustainability attributes as they did not fatten their calves and therefore their productivity decreases. This could be an important weakness as there is no market for weaned calves and, as a consequence, farmers depend upon intermediaries (traders and feedlots) for selling their calves.

On the contrary, Organic 2 farms are more dependent on external resources, which can lead to a high reduction of their sustainability if an increase of concentrates' prices occurs. However, although consumers' demand for organic beef still remains low, Organic 2 are predictably less vulnerable, due to the fact that they have established contracts with retailers, which reduce the importance of such low self-reliance and their level of risk. With respect to the dependence on subsidies, none of the three production systems seems to be less dependent on them than the others. However, organic ruminants systems located in other areas have been found to be highly dependent on subsidies (Hrabalová and Zander 2006, Argyropoulos et al. 2013). This divergence on the results can be explained by the fact that, as Lobley et al. (2009) and Lobley et al. (2013) pointed out, the differences among organic and conventional farms are mainly due to their structure and management, more than to the fact of being organic or not.

From the equity point of view, no major differences were found among systems. However, the indices related to the workforce showed interesting differences that have already been addressed. According to Morison et al. (2005), organic farms showed less mean jobs per area (calculated as "Total AWU per 100 ha UAA in our study). However, comparisons among different studies must be interpreted carefully, since job creation varies greatly with enterprise type, farm size and regions (Morison et al. 2005, Lobley et al. 2009).

Regarding the distance to the slaughterhouse, the differences found are worthy of a deeper analysis. As it has been mentioned above, organic farms are further from the slaughterhouses, which is due to the low number of organically certified slaughtering premises. In the case that demand for organic beef would increase, the development of the organic beef sector would be difficult since this scarce development of the organic industry. Additionally, this greater distance makes it difficult to implement activities of elaboration and direct sales of their products to consumers, which would allow the producer to charge a price premium, which is, according to (Tzouramani et al. 2011), indispensable for organic production because of the higher costs involved.

With regard to the stability and resilience of the farms, it has been observed that organic farms showed a reduced use in synthetic inputs (pesticides, mineral fertilizers, and preventive antiparasitics and antibiotics). It had been observed previously in beef cattle farms in Spain (Blanco-Penedo et al. 2012) and, according to Anderson et al. (2012) and Sanderson et al. (2013), has positive environmental and economic effects. Moreover, Organic 2 farms had a greater degree of integration of different livestock species, crops, and wooded areas. This integration allows the farms to make better use of their food resources, to conserve their natural resources and landscape diversity, and strengthen their ecosystem services, while providing themselves with their own agricultural inputs, and reducing their external dependence and vulnerability. Additionally, it is interesting to study the higher scores for carbon sequestration of the farms analyzed, since enhancing carbon sequestration is the main approach that rangelands and agroforestry systems can adopt to address climate change. In fact, rangelands have a great potential for storing carbon (Schuman et al. 2002). In such systems, improved land management practices and the presence of trees contribute to carbon sequestration, thereby improving soil quality (Nair et al. 2009) and mitigating climate change.

Moreover, their use of native species better adapted to local conditions, means that livestock productivity is less affected by extreme weather conditions (such as the high summer temperatures. This aspect is important due to the climate changes that Spain is predicted to undergo (Segnalini et al. 2013). For all these reasons, the Organic 2 system is thus expected to be more stable.

In addition, it is important to say that environmentally friendly production systems may not lead to long-term economic and social sustainability of a farm (Darnhofer et al. 2010). However, organic and resource-conserving agriculture may help smallholders (as those of dehesa) to navigate changing environmental and market conditions (Bennett and Franzel 2013). Moreover, the interaction between the social, ecological and economic dimensions are of great importance in the resilience thinking (Darnhofer et al. 2010).

Finally, and with respect to the Productivity, Organic 2 farms again obtained the highest scores. The results shed light on this issue, since few comparisons have been made regarding productivity and profitability between organic and conventional beef cattle farms. Those carried out not always compare to two systems under the same conditions (two extensive systems: organic vs. Conventional). Thus, the main differences found in other studies come from the structure and management of the farms, rather than whether the farms are organic or conventional. Nevertheless, some conclusions can be made from those results. It seems that organic cattle farms tend to be less productive and/or less profitable than conventional farms due to several reasons: longer productive period, overhead costs specially the feedstuffs, and lower revenues (Blanco-Penedo et al. 2012, Gillespie and Nehring 2013). In the present study, no differences were found between the organic farms that sold organic products (Organic 2 farms) and the Conventional ones.

## Conclusion

This indicator-based comparative evaluation of the sustainability of organic and conventional beef cattle farms has allowed us to establish levels of sustainability in technical, environmental, social, and economic terms for these types of production systems in a high-value Mediterranean agro-ecosystem. Regarding sustainability attributes considered in the study, Organic 2 farms had the best scores on the Adaptability and Flexibility, Stability, and Productivity attributes. Organic 1 farms had the best scores on the Self-reliance attribute, due mainly to not fattening their calves.

In terms of overall sustainability, farms belonging to the Organic 2 system were found to be the most sustainable. These farms are in an advantageous position for their present and future sustainability in view of the orientations of the new Common Agricultural Policy (environmentally friendly production systems), as well as the trend in consumer demand for products that are more natural, healthy, and environmentally sustainable. In summary the organic, complete cycle, beef cattle production systems is a more sustainable option than the other two under these indicators. Nonetheless, implementing these systems is complex due to the high costs involved in the production of organic calves and to the weak demand for them.

In general, all three types of farm need to improve in certain common aspects that are crucial in the current and future context of the livestock sector. These aspects are: reducing the dependence on external feed, implementing more environmentally friendly farming practices and farm diversification. The result will be both to lessen the farm's vulnerability and to increase its production of environmental and social services. Indeed, it is in these latter where the dehesa beef cattle farms can stand out and benefit from the market and from agricultural policies.

## Acknowledgments

This work was supported by the Spanish Ministry of the Economy and Competitiveness under Grant [INIA-RTA2009-00122-C03-03].

## References

Anderson, D.M., Fredrickson, E.L., Estell, R.E.(2012). Managing livestock using animal behaviour: mixed-species stocking and flerds. *Animal*, 6 (8), 1339-1349.

Argyropoulos, C., Tsiafouli, M.A., Sgardelis, S.P., Pantis, J.D. (2013). Organic farming without organic products. *Land Use Policy*, 32, 324-328.

Astier, M., García-Barrios, L., Galván-Miyoshi, Y., González-Esquivel, C.E.,Masera, O.R.(2012). Assessing the Sustainability of Small Farmer Natural Resource Management Systems. A Critical Analysis of the MESMIS Program (1995-2010). *Ecology and Society*, 17 (3) 25.

Bennett, M., Franzel, S.(2013). Can organic and resource-conserving agriculture improve livelihoods? A synthesis. *International Journal of Agricultural Sustainability*, 11(), 193-215.

Blanco-Penedo, I., López-Alonso, M., Shore, R.F., Miranda, M., Castillo, C., Hernández, J., Benedito, J.L.(2012).Evaluation of organic, conventional and intensive beef farm systems: health, management and animal production. *Animal*, 6 (9), 1503-1511.

Coelho, I.S.(1994). Economia do Montado. Análise Económica de Tres Montados de Sobro Alentejanos. *Silva Lusitana*, 2 (2): 133-141.

Council Regulation (EC) No 834/2007 of 28 June 2007 on organic production and labelling of organic products and repealing Regulation (EEC) No 2092/91.

Dale, V.H., Efroymson, R.A., Kline, K.L., Langholtz, M.H., Leiby, P.N., Oladosu, G.A., Davis, M.R., Downing, M.E., Hilliard, M.R.(2013). Indicators for assessing socioeconomic sustainability of bioenergy systems: A short list of practical measures. *Ecological Indicators*. 26, 87-102.

Darnhofer, I., Fairweather, H., Moller, H.(2010). Assessing a farm's sustainability: insights from resilience thinking. *International Journal of Agricultural Sustainability*. 8 (), 186-198.

EAA.(2001). European Communities. Manual on the economics accounts for agriculture and forestry EAA/EAF 97 (Rev 1.1). Office for Official Publications of the European Communities, Luxembourg.

Gaspar, P., Mesías, F. J., Escribano, M., Rodriguez de Ledesma, A., Pulido, F.(2007).Economic and management characterization of dehesa farms: implications for their sustainability. *Agroforestry Systems*, 71 (3), 151-162.

Gaspar P., Escribano M., Mesías F.J., Rodríguez de Ledesma A., Pulido F.(2008). Sheep farms in the Spanish rangelands (dehesas): Typologies according to livestock management and economic indicators. *Small Ruminant Research*, 74 (1-3), 52-63.

Gaspar, P., Mesías, F.J., Escribano, M., Pulido, F. (2009a).Sustainability in Spanish extensive farms (dehesas): an economic and management indicator-based evaluation. *Range land Ecology & Management*, 62 (2), 153-162.

Gaspar, P., Mesías, F.J., Escribano, M., Pulido, F.(2009b). Evaluación de la sostenibilidad en explotaciones de dehesa en función de su tamaño productivo y orientación ganadera. ITEA, *Información Técnica Económica Agraria*, 105 (2), 117-141.

Gillespie, J., Nehring, R.(2013). Comparing economic performance of organic and conventional U.S. beef farms using matching samples. *Australian Journal of Agricultural and Resource Economics*, 57 (2), 178–192.

Gómez-Limón, J.A., Sánchez-Fernández, G.(2010). Empirical evaluation of agricultural sustainability using composite indicators. *Ecological Economics*, 69 (5), 1062-1075.

Halberg, N.(2012). Assessment of the environmental sustainability of organic farming: Definitions, indicators and the major challenges. *Canadian Journal of Plant Science*, 92 (6), 981-996.

Hernández, C.G.(1998). Ecología y fisiología de la dehesas. I n: C.G. Hernández, ed.), La dehesa, aprovechamiento sostenible de los recursos naturales. *Editorial Agrícola Española*, Madrid: 53-94.

Hrabalová, A., Zander, K.(2006). Organic beef farming in the Czech Republic: structure, development and economic performance. *Agricultural Economics* CZECH, 52 (2), 89-100.

Lebacq, T., Baret, P.V., Stilmant, D.(2013).Sustainability indicators for livestock farming.A review. *Agronomy for Sustainable Development*, 33 (2), 311-327.

Lobley, M., Butler, A., Reed, M.(2009). The contribution of organic farming to rural development: An exploration of the socio-economic linkages of organic and non-organic farms in England. *Land Use Policy*, 26 (3), 723-735.

Lobley, M., Butler, A., Winter, M.(2013). Local Organic Food for Local People?Organic Marketing Strategies in England and Wales. Regional Studies, 47 (2), 216-228. doi: 10.1080/00343404.2010.546780.

López-Ridaura, S., Masera, O.R., Astier, M.(2002). Evaluating the sustainability of complex socio-environmental systems: the MESMIS framework. *Ecological Indicators*, 2 (1-2), 135-148.

MAGRAMA.(2007). Tercer Inventario Nacional Forestal [online]. Ministerio de Agricultura, Alimentación y Medio Ambiente. Available from: http://www.magrama.gob.es/es/biodiversidad/temas/inventarios-nacionales/inventario-forestal-nacional/default.aspx. [Accessed 26 November 2013].

Masera, O., Astier, S., López-Ridaura, S.(1999). Sustentabilidad y manejo de los recursos naturales. El marco de evaluación MESMIS. Gira, IE-UNAM, México: Mundi-Prensa, S.A.

Milán, M.J., Bartolomé, J., Quintanilla, R., García-Cachán, M.D., Espejo, M., Herraiz, P.L., Sánchez-Recio, J.M., Piedrafita, J.(2006). Structural characterisation and typology of beef cattle farms of Spanish wooded rangelands (dehesas). *Livestock Science*, 99 (2-3), 197–209.

Montero, G., San Miguel, A., Cañellas, I.(1998). Sistemas de selvicultura mediterránea. La dehesa. En: Jiménez R.M. y Lamo de Espinosa, J. (eds.) AgriculturaSostenible. Mundi-Prensa, Madrid, pp. 519-554.

Morison, J., Hine, R., Pretty, J.(2005). Survey and analysis of labour on organic farms in the UK and Republic of Ireland. *International Journal of Agricultural Sustainability*, 3 (1), 23-43.

Murillo, L.R., Corrales, N., Escribano, M., Sánchez, M.(2009). Masas arboladas y fijación de emisiones. In: L.R. Murillo, ed. Balanzas ambientales de España. Cáceres: Servicio de publicaciones de la Universidad de Extremadura, 77-104.

Nahed, J., Castel, J.M., Mena, Y., Caravaca, F.(2006). Appraisal of the sustainability of dairy goat systems in Southern Spain according to their degree of intensification. *Livestock Science*, 101 (1-3), 10–23.

Nair, P.K.R., Kumar, B.M., Nair, V.D.(2009). Agroforestry as a strategy for carbon sequestration. *Journal of Plant Nutrition and Soil Science*, 172 (1), 10-23. doi: 10.1002/jpln.200800030.

Nardone, A., Ronchi, B., Lacetera, N., Ranieri, M.S., Bernabucci, U.(2010). Effects of climate changes on animal production and sustainability of livestock systems. *Livestock Science*, 130 (1-3),57–69.

Rasul, G., Thapa, G.B.(2004). Sustainability of ecological and conventional agricultural systems in Bangladesh: an assessment based on environmental, economic and social perspectives. *Agricultural Systems*, 79 (3), 327-351.

Regulation (EC) No 138/2004 of the European Parliament and of the Council of 5 December 2003 on the economic accounts for agriculture in the Community.

Ripoll-Bosch, R., Díez-Unquera, B., Ruiz, R., Villalba, D., Molina, E., Joy, M., Olaizola, A., Bernués, A.(2012). An integrated sustainability assessment of Mediterranean sheep farms with different degrees of intensification. *Agricultural Systems*, 105 (1), 46-56.

Ripoll-Bosch, R., Joy, M., Bernués, A.(2013). Role of self-sufficiency, productivity and diversification on the economic sustainability of farming systems with autochthonous sheep breeds in less favoured areas in Southern Europe. *Animal* 4 (1-9), 1-9.

Sanderson, M.A., Archer, D., Hendrickson, J., Kronberg, S., Liebig, M., Nichols, K., Schmer, M., Tanaka, D., Aguilar, J.(2013). Diversification and ecosystem services for conservation agriculture: outcomes from pastures and integrated crop-livestock systems. *Renewable Agriculture and Food Systems*, 28 (2), 129-144.

Schuman, G.E., Janzen, H.H., Herrick, J.E.(2002). Soil carbon dynamics and potential carbon sequestration by rangelands. *Environmental Pollution*, 116 (3), 391-396.

Segnalini, M., Bernabucci, U., Vitali, A., Nardone, A., Lacetera, N.(2013).Temperature humidity index scenarios in the Mediterranean basin. *International Journal of Biotemeorology*, 57 (3), 451-458.

Serrano, E.P., Lavín, P., Giradles, F.J., Bernués, A., Ruiz, A.(2004). Classification variables of cattle farms in the mountains of León, Spain. *Spanish Journal of Agricultural Research*, 2(4), 504–511.

Smith, J., Pearce, B.D., Wolfe, M.S.(2011). Reconciling productivity with protection of the environment: Is temperate agroforestry the answer? *Renewable Agriculture and Food Systems*, 28 (1), 80-92.

Schnabel, S. (1997). Soil erosion and runoff production in a small watershed under silvo-pastoral landuse (Dehesa) in Extremadura, Spain.Geoforma Ediciones. *Logroño*, 167 pp.

Ten Brink, B.J.E., Hosper, S.H., Colin, F.(1991).A quantitative method for description and assessment of ecosystems: the AMOEBA approach. *Marine Pollution Bulletin*, 23, 265-270.

Tuomisto, H.L., Hodge, I.D., Riordan, P., Macdonald, D.W.(2012). Does organic farming reduce environmental impacts? - A meta-analysis of European research. *Journal of Environmental Management*, 112 (15), 309-320.

Tzouramani, I., Sintori, A., Liontakis, A., Karanikolas, P., Alexopoulos, G.(2011). An assessment of the economic performance of organic dairy sheep farming in Greece. *Livestock Science*, 141 (2-3), 136–142.

In: Rangeland Ecology, Management ...
Editor: Victor R. Squires
ISBN: 978-1-63482-504-7


*Chapter 5*

# CHANGES IN VEGETATION AND IN THE ORGANIC SOIL FRACTIONS RELATED WITH GRAZING AND FIRE HISTORIES IN THE PATAGONIAN MONTE, ARGENTINA

***Guadalupe Peter[1,3,*], Flavia Alejandra Funk[2], Martín Alejandro Luna[1,3] and Silvia Susana Torres Robles[1]***
[1]Sede Atlántica, Universidad Nacional de Río Negro, Viedma, Argentina
[2]Centro de Recursos Naturales Renovables de la Zona Semiárida-CONICET, Bahía Blanca, Argentina
[3]CONICET, Bahía Blanca, Argentina

## ABSTRACT

In terrestrial ecosystems the vegetation and soil co-evoluted in response to some disturbance. In arid and semi-arid lands around the world vegetation is distributed in patches within a bare-soil matrix. Such vegetation and soil are more vulnerable to disturbance effects and highly susceptible to desertification, vegetation attributes and organic fractions of the soil are the parameters widely used for evaluating changes in ecosystems. In the North-east Patagonian Monte in Argentina, the vegetation is a shrubland steppe with patches dominated by shrubs, with grasses, forbs and cryptogams under their canopy. It was hypothesized that grazing increases patchiness across its associated trampling and selective foraging; and fires and posterior wind erosion homogenize the distribution of soil particles and vegetation with consequence on dynamic of soil organic carbon. It was predicted that there would be: 1) greater cover, specific frequency and richness of shrubs in grazed sites; 2) greater cover, specific frequency and richness of herbs and preferred grasses in ungrazed sites; 3) a random pattern of vegetation distribution in burnt areas; and 4) greater carbon soil stock in ungrazed sites and lower in burnt ones. Aerial cover of all perennial species, distribution

* E-mail: gpeter@unrn.edu.ar.

of soil particles sizes, and stocked organic carbon (SOC) of soil were measured at six sites with different land-use histories: heavily grazed, medium grazed, ungrazed, long exclosure from grazing followed by grazing, burnt and then ungrazed, and burnt and then grazed. Species were grouped into five functional types: shrubs, sub-shrubs, preferred grasses, non-preferred grasses and forbs. The results showed significant differences in the cover of preferred and non-preferred grasses, forbs and total cover with previous grazing but there was no evidence of shrub encroachment. Species frequency and richness decreased especially with increased grazing intensity. The pattern of spatial distribution changed from aggregated in grazed sites to random in ungrazed and burnt sites for all plant functional types. Land use history affects mainly the distribution of particle sizes in the soil and the bulk density. We did not find differences in total SOC, but SOC in interpatches of highly grazed site is associated with mineral particles, whereas in patches of medium grazed is greater in the coarse fraction. The content of carbon was not augment by fire. It is concluded that moderate disturbance (fire followed by grazing or medium grazing intensity) increased the plant diversity and the amount of rapidly available carbon in the soil.

**Keywords:** grazing intensity, stocked soil carbon, disturbance, exclosure, plant functional types, distribution pattern

## INTRODUCTION

In arid and semi-arid lands around the world, vegetation is distributed in patches within a bare-soil matrix (Aguiar and Sala 1999); these vegetation patches have a greater concentration of sediments, nutrients, seed and water, and are denominated phytogenic mounds. In most arid dry lands, overgrazing by domestic livestock is the principal cause of desertification (Fleischner 1994; Whitford et al. 1997; Allington and Valone 2011) and of shrub encroachment (Puigdefábregas 1998; Roques et al. 2001; Parizek et al. 2002; Villagra et al. 2009).

Overgrazing can cause an increase in the cover of annual and shrub species and a decrease in the cover of perennial grasses (Milchunas and Lauenroth 1993; Bertiller and Bisigato 1998). These changes in cover are reflected in the structure of the vegetation patches. Thus, in areas with high grazing pressure, patches dominated by preferred grasses disappear and are replaced by shrub species with grasses under their canopy (Bertiller and Bisigato 1998; Aguiar and Sala 1999). This can lead to changes in the structure of the plant community from gramineous steppe to shrubland. The magnitude of the grazing impact on these systems depends on the type of herbivore and grazing management (for example, see Bisigato and Bertiller 1997; Bertiller and Bisigato 1998; Olff and Ritchie 1998; Bertiller et al. 2002; Cingolani et al. 2005; Loydi et al. 2012).

These changes in the structure, also change important processes such as nutrients cycling and water flux (Reynolds et al. 1999). In the interpatch spaces with bare soil, the soil erosion from wind and water cause losses of nutrients, which are accumulated under shrubs. The nutrient enriched sediment formed “islands of fertility” (Reynolds et al. 1999; Ravi et al. 2009). Islands of fertility have higher concentrations of nitrogen, phosphorus and carbon under shrub canopies, compared to adjacent interpatch spaces (Allington and Valone 2014).

Therefore, the low availability of nutrients in the interpatches limits the recovery of grasses (Schlesinger et al. 1990; D’Odorico et al. 2007) leading to degradation of the systems.

Also, livestock grazing enhances the formation of this pattern of fertile islands (Allington and Valone 2014).

The soil organic carbon (SOC) is an important indicator of soil and land health as it integrates several inherent soil properties. The total amount of organic carbon stocked in a soil is the resulting net balance of all carbon fluxes entering and leaving the soil over time. Most organic matter of the soil derives from plants (above and below ground) (Kutsch et al. 2009). SOC responds strongly to aboveground landscape dynamics, including land-use changes and land degradation processes (Galantini et al. 2008; Duval et al. 2013; Toledo et al. 2013).

The reduction of the vegetal biomass, generated by overgrazing and fire events, is the principal factor that reduce the organic carbon stocked in the soil (FAO 2002; Gaitán et al. 2009). SOC and its different fractions are soil parameters highly influenced by management practices, mainly the young or labile SOC fraction, which is important to interpret changes in soil quality (Kapkiyai et al. 1999; Galantini et al. 2008; Duval et al. 2013; Toledo et al. 2013), being a more sensitive indicator than SOC and total nitrogen (Nt) (Biederbeck et al. 1994).

However, the physical conditions of soil and climate must be considered as factors that have a strong influence on the labile carbon fractions (Galantini and Rosell 2006; Duval et al. 2013).

Shrub encroachment has been recognized as one of the most important ecological and economic issues in semi-arid rangelands (Rostagno et al. 2006). The fire can be a useful tool to reverse shrub encroachment and to promote the recovery of grasses because post-fire wind-erosion processes redistribute nutrient resources accumulated by the shrubs in the phytogenic mounds and, hence, change the spatial patterns of vegetation (Hodgkinson and Harrington 1985; West and Hassan 1985; Wambolt and Payne 1986; Pfeiffer and Steuter 1994; Bóo et al. 1996; Rostagno et al. 2006; Ravi et al. 2007; Peláez et al. 2010). Fire could produce lost of soil organic carbon because its volatilization during the fire (Gaitán et al. 2007). The magnitude of this loss depends on the frequency and intensity of fire (Oesterheld et al. 1999). In accordance with Ravi et al. (2006), the organic compounds are volatilized by the fire and transported into the soil by the strong temperature gradients existing through the soil profile. This condensation results in the hidrophobicity of the soil that increases soil erosion post fire. Moreover, it has been suggested that burning should be used in combination with grazing management, as high grazing rates can reduce grass cover and also facilitate enhanced shrub seed dispersal (Ravi et al. 2009).

In the semi-arid province of Monte in the North-east Patagonian region of Argentina, the vegetation is distributed in patches dominated by shrubs, with grasses, forbs and cryptogams under their canopy (Bisigato and Bertiller 1997; Busso et al. 2012; Kröpfl et al. 2013). These vegetation patches are in mounds that act as fertility islands (Gaitán et al. 2007). Overgrazing in the Patagonian rangelands can cause a reduction in plant cover (Bisigato and Bertiller 1997, Peter et al. 2013), changes in botanical composition (Bertiller and Bisigato 1998; Peter et al. 2012, 2013), increase in shrub cover (Bertiller and Bisigato 1998), and reduction or even suppression of fires (Peláez et al. 2003; Cingolani et al. 2005; Cingolani et al. 2008; Tizón et al. 2010; Loydi et al. 2012).

It was hypothesized that grazing increases patchiness across its associated trampling and selective foraging; and fires and posterior wind erosion homogenize the distribution of soil particles and vegetation with consequence on dynamic of organic fractions of soil. The present study was designed to test four predictions, arising from this hypothesis, namely that:

1) greater cover, specific frequency and richness of herbs and preferred grasses in ungrazed sites; 2) greater cover, specific frequency and richness of shrubs in grazed sites; 3) a random pattern of vegetation distribution in burnt areas; and 4) greater stocked soil carbon in ungrazed sites, and lower in burnt ones.

# METHODS

## Study Area

The study area was located in North-east Patagonia, in the Departments of Adolfo Alsina and San Antonio in Río Negro, Argentina (40° 40' S, 64° 10' W). Phytogeographically, it belongs to the Monte province, Southern District, Northern Patagonic sub-district (Roig et al. 2009). The area is characterized by a transition sub-temperate semi-arid climate, with warm summers (mean temperature 21°C) and mild winters (mean temperature 7°C). Average annual precipitation is between 200 and 350 mm, with high intra-annual variability (Bran et al. 2000). The climate is windy, especially in spring and summer, which contributes to the soil moisture deficit. Soils are aridisoils and entisoils, classified suchustolic Natrargides on the plains and fluventic Haplocambides on hillocks according to soil taxonomy 1996 (USDA 1996; Godagnone and Bran 2009).

Vegetation is a shrubland steppe, with an herbaceous stratum of predominantly winter-growing grasses. The dominant shrub species are *Chuquiraga erinacea* D. Don, *Larrea divaricata* Cav., *L. cuneifolia* Cav., *Condalia microphylla* Cav. and *Prosopis alpataco* Phil. Among the dominant herbaceous species are *Nassella tenuis* (Phil.) Barkworth, *Poa ligularis* Nees. ex Steud.,*Jarava plumosa* (Spreng.) S.W.L. Jacobs and J. Everett, *Pappostipa speciosa* (Trin. and Rupr.) Romasch.,*Schismus barbatus* (L.) Thell. and *Erodium cicutarium* (L.) L'Hér. ex Aiton (Cabrera 1971; Godagnone and Bran 2009).

The principal production activities in the area are the extensive grazing and firewood extraction (Arturi et al. 2011). The study site had been grazed by sheep for almost 100 years with a normal annual stocking rate in the region of ~0.18 sheep $ha^{-1}$. Natural fires are frequent in the area (Gaitán et al. 2007), especially in the humid years, when the water availability promotes growing and biomass accumulation. Then, dry vegetal material turned on fire in summer because of high temperatures and winds widespread the fire focus.

## Sites and Field Sampling

The topography in the study area was uniformly flat and is < 100 m above sea level (Godagnone and Bran 2009). The area is grazed by native herbivores - the guanaco (*Lama guanicoe*) is the most abundant species – and, as we mentioned above, domestic herbivores have been introduced for about 100 years. The area of study was 30 x 10 km of a vegetation unit which comprised several ranches with different land uses. The vegetation unit was defined as a medium shrubby steppe, with some specific variation in relation with the soil texture (Godagnone and Bran 2009).

A railway passes through the area, and an area of land 200 m wide along the railway was protected from grazing by fences, which provided an exclosure of 200 m x 30 km (600 ha). Further, after a road was constructed, an area of 100 ha was isolated from grazing for 40 years because there was no permanent water point. It was then grazed by sheep at the normal stocking rate of ~0.18 sheep $ha^{-1}$ for one year.

Eight years before sampling, wild fires partially affected the study area. This fire event allowed the study of the role of grazing and fire on the vegetation community, because after the fire an area of 1 ha was excluded from grazing by fences.

Piosphere formation is common around permanent water points in arid areas (Bisigato et al. 2005), so a grazing intensity gradient, based on the distance from the water point, was studied.

Six sites were selected with different land-use histories (Table 1). Two sites with different grazing intensities due to distance from water were selected: $\leq$ 600 m from the water point (heavily grazed area, site HG), $\geq$ 2500 m from the water point (medium grazed area, site MG). A third site was located near the railway from which domestic livestock had been excluded for a long time (ungrazed area, site UG). A fourth site was located in the 100 ha exclosure which was grazed by sheep for one year after 40 years of exclusion of (long ungrazed-grazed area, site LU). The final two sites were selected in burnt places: one in the 1 ha area in which livestock was excluded after fire (burnt ungrazed area, site BU); and the other in an adjacent site which remained grazed by sheep at the normal stocking rate (burnt grazed area, site BG).

We located 30 plots within the six land-use sites in the study area. Then, at each site, five plots of 10 x 10 m were randomly established. This plot number and size was determined from preliminary studies using the methods of Mueller-Dombois and Ellenberg (1974) and Mateucci and Colma (1982). In the spring of 2010 the aerial cover of all perennial species on each plot at all six sites was measured following the Braun-Blanquet method (Braun-Blanquet 1972). Species were grouped into five functional types: shrubs, sub-shrubs, preferred grasses, non-preferred grasses and forbs.

**Table 1. Description of the land use history of the six sites**

| Land use history | Description of the site |
|---|---|
| Medium grazed (MG) | Grazed area with typical stocking rate for the region |
| High grazed (HG) | Highly grazed area at $\leq$ 600 m of the water point |
| Ungrazed (UG) | Exclosures generated by fencing of a railroad |
| Long ungrazed-grazed (LU) | Area excluded from grazing for 40 years of which was grazed the year before the data collection |
| Burnt-ungrazed (BU) | Area which was burnt in a wildfire, excluded for ten years after the fire |
| Burnt-grazed (BG) | Area which was burnt in a wildfire, continuously grazed for ten years after the fire |

## Soil Sampling

In each plot were taken undisturbed samples of soil, six in the intershrub spaces (interpatch) and six under shrub canopies (patch), with cylinder of 5 cm in height and 4.7 cm in diameter in order to calculate the bulk density (BD) and makes a soil size fractionation.

## Soil Physical and Chemical Analyses

Half of soil samples were dried in oven at 105 °C until constant weight. Soil BD was then calculated by dividing the total dry weight by the soil core volume (Blake and Hartge 1986).

The remaining samples were air dried and passed through one sieve of 2000 μm mesh. Each sample was divided in two subsamples. One subsample was passed through one sieve of 500 μm meshes and used to determinate the total concentration of soil organic carbon (SOC). The other subsample was used to determinate the size fractionation of SOC with the wet sieving of soil method (Duval et al. 2013). The sieving was done with a pair of sieves of 53 μm and 105 μm of diameter mesh. Then, three fractions of soil were obtained: CF= the coarse fraction (105–2000 μm), which included coarse particulate organic carbon and fine to coarse sands; MF= the medium fraction (53–105 μm) which included fine particulate organic carbon and very fine sand; and FF= the fine fraction (<53 μm), containing mineral associated organic carbon as well as silt and clay minerals. The concentration of organic carbon of all fractions obtained was determined by wet combustion (Walkley and Black 1934).

The stocked of organic carbon, total (SOC) and for each fraction of soil ($SOC_{CF}$, $SOC_{MF}$ and $SOC_{FF}$), was quantified using the following equation (Ussiri et al. 2006):

$$SOC\ (t\ ha^{-1}) = (X/100)*BD*d*10^4\ m^2\ ha^{-1}$$

where X is the concentration of organic carbon (%), BD is bulk density (t $m^{-3}$) and *d* is the soil thickness (m).

## Statistical Analyses

*Vegetation attributes.* The similarity between sites was determined using Principal Component Analysis (PCA) (Digby and Kempton 1987). The similarity with respect to species frequency (number of plots where a species was present in each site), species richness (number of species of each plant functional type in each site) and aerial cover ($m^2$) of each species and of each plant functional type in each site were analyzed. The data on vegetation cover (whether transformed or untransformed) did not follow a normal distribution (Shapiro-Wilks Normality Test) so the non-parametric Kruskal–Wallis test for multiple comparisons between sites was applied with a post-hoc paired comparison test (Balzarini et al. 2008).

*Spatial distribution.* To analyze species pattern, the aerial cover of each species was used. For functional group patterns, the cover of all species in each functional group in the plot was summed. For all six land-use sites the variance/mean for each species and functional type was calculated (Mateucci and Colma 1982), testing the significance with a $\chi^2$ test (de la Cruz Rot 2008). If the variance/mean (n-1) was lower than the critical value ($\chi^2_{(0.025;4)}$=0.484), the distribution pattern was considered to be regular; if the rate was greater than the critical value ($\chi^2_{(0.975;4)}$=11.14), the distribution pattern was classed as aggregated; and if the rate was between the critical values (0.484 and 11.14) then the pattern was considered to be random (de la Cruz Rot 2008).

*Soil properties.* The similarity between sites was determined using PCA with respect to SOC, CF (weight of particles between 105-2000 μm), MF (weight of particles between 53-105 μm), FF (weight of particles <53 μm), $SOC_{CF}$, $SOC_{MF}$ and $SOC_{FF}$. All data were analyzed using one-way ANOVA. The variables (interpatches: $SOC_{FF}$, patches: $SOC_{MF}$, SOC) which not follow a normal distribution (Shapiro-Wilks Normality Test) were previously transformed to natural logarithm, to meet normality requirements. Means were compared using Fisher LSD tests.

All the analyses were undertaken using InfoStat software (Di Rienzo et al. 2011).

# RESULTS

## Vegetation Attributes

Based on species frequency, PC1 of the Principal Components Analysis explained almost 42 % of the variability and PC2 27 %. Sites with different land-use histories were ordered in a disturbance gradient on PC1, principally by grazing intensity (Figure 1).

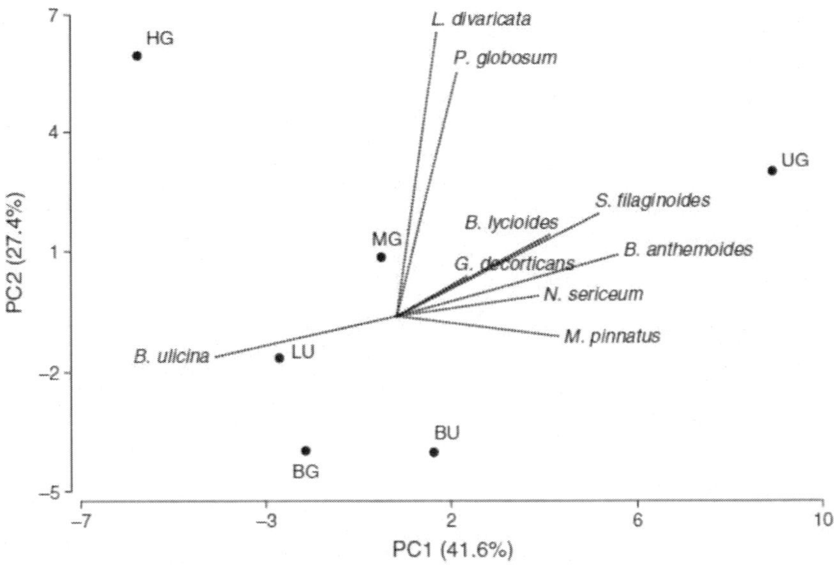


Figure 1. Principal Components Analysis showing the relationships of similarities between the six sites with respect to species frequency. Only species with correlations greater than 0.80 are indicated. The six sites were: Ungrazed site (UG), Medium grazed site (MG), Heavily grazed site (HG), Ungrazed for a long period then grazed site (LU), Burnt then ungrazed site (BU), and Burnt and then grazed (BG).

Modified from Peter et al. 2013.

**Table 2. Changes in species frequency (number of plots where the species was present) in sites with different land use histories**

| Species | Situations | | | | | |
|---|---|---|---|---|---|---|
| **Shrubs** | BG | BU | MG | UG | LU | HG |
| *Condalia microphylla* | 4-5 | 4-5 | 4-5 | 4-5 | 4-5 | 4-5 |
| *Lycium* spp. | 4-5 | 4-5 | 4-5 | 4-5 | 4-5 | 4-5 |
| *Prosopis alpataco* | 4-5 | 4-5 | 4-5 | 4-5 | 4-5 | 4-5 |
| *Schinus johnstonii* | 4-5 | 4-5 | 4-5 | 4-5 | 2-3 | 2-3 |
| *Chuquiraga erinacea* | 4-5 | 4-5 | 4-5 | 4-5 | 4-5 | ≤1 |
| *Larrea divaricata* | ≤1 | ≤1 | 4-5 | 4-5 | 4-5 | 4-5 |
| *Prosopidastrum globosum* | ≤1 | ≤1 | ≤1 | 4-5 | ≤1 | 4-5 |
| *Brachyclados lycioides* | ≤1 | ≤1 | ≤1 | 2-3 | ≤1 | ≤1 |
| *Ephedra ochreata* | ≤1 | ≤1 | ≤1 | 2-3 | ≤1 | ≤1 |
| *Bredemeyera microphylla* | ≤1 | 2-3 | ≤1 | ≤1 | ≤1 | ≤1 |
| *Cyclolepis genistoides* | ≤1 | ≤1 | ≤1 | ≤1 | ≤1 | ≤1 |
| *Geoffroea decorticans* | ≤1 | ≤1 | ≤1 | ≤1 | ≤1 | ≤1 |
| *Mulguraea ligustrina* | ≤1 | ≤1 | ≤1 | ≤1 | ≤1 | ≤1 |
| *Larrea cuneifolia* | ≤1 | ≤1 | ≤1 | ≤1 | ≤1 | ≤1 |
| *Monttea aphylla* | ≤1 | ≤1 | ≤1 | ≤1 | ≤1 | ≤1 |
| **Subshrubs** | | | | | | |
| *Margyricarpus pinnatus* | ≤1 | 2-3 | ≤1 | 4-5 | 2-3 | ≤1 |
| *Senna aphylla* | 4-5 | 4-5 | 4-5 | 4-5 | 4-5 | 2-3 |
| *Baccharis ulicina* | 4-5 | 4-5 | 4-5 | ≤1 | 4-5 | 4-5 |
| *Baccharis trimera* | ≤1 | 2-3 | 2-3 | 2-3 | ≤1 | ≤1 |
| *Ephedra triandra* | 2-3 | ≤1 | ≤1 | ≤1 | 4-5 | ≤1 |
| *Grindelia chiloensis* | ≤1 | ≤1 | ≤1 | ≤1 | 2-3 | ≤1 |
| *Sphaeralcea mendocina* | ≤1 | ≤1 | ≤1 | ≤1 | 4-5 | ≤1 |
| *Acantholippia seriphioides* | ≤1 | ≤1 | ≤1 | ≤1 | ≤1 | ≤1 |
| *Baccharis crispa* | ≤1 | ≤1 | ≤1 | ≤1 | ≤1 | ≤1 |
| **Non palatable grasses** | | | | | | |
| *Pappostipa speciosa* | 4-5 | 4-5 | 4-5 | ≤1 | 4-5 | 2-3 |
| *Melica rigida* | ≤1 | ≤1 | ≤1 | ≤1 | ≤1 | ≤1 |
| **Palatable grasses** | | | | | | |
| *Poa ligularis* | 4-5 | 4-5 | 4-5 | 4-5 | 2-3 | 2-3 |
| *Poa lanuginosa* | 2-3 | 4-5 | 4-5 | 2-3 | 2-3 | 4-5 |
| *Nassella tenuis* | ≤1 | ≤1 | ≤1 | ≤1 | 2-3 | 2-3 |
| *Jarava plumosa* | ≤1 | ≤1 | ≤1 | ≤1 | 2-3 | ≤1 |
| *Pappophorum caespitosum* | 2-3 | 2-3 | ≤1 | ≤1 | ≤1 | ≤1 |
| *Jarava neaei* | ≤1 | ≤1 | ≤1 | ≤1 | ≤1 | ≤1 |
| *Piptochaetium napostaense* | ≤1 | ≤1 | ≤1 | ≤1 | ≤1 | ≤1 |
| **Forbs** | | | | | | |
| *Senecio filaginoides* | ≤1 | ≤1 | ≤1 | 4-5 | ≤1 | ≤1 |
| *Baccharis gilliesii* | 2-3 | 4-5 | 2-3 | 4-5 | 2-3 | ≤1 |
| *Boopis anthemoides* | ≤1 | 2-3 | ≤1 | 4-5 | 2-3 | ≤1 |
| *Senecio melanopotamicum* | ≤1 | 2-3 | 2-3 | 4-5 | ≤1 | ≤1 |
| *Baccharis melanopotamica* | 2-3 | ≤1 | ≤1 | 2-3 | 2-3 | ≤1 |
| *Noticastrum sericeum* | ≤1 | ≤1 | ≤1 | 2-3 | ≤1 | ≤1 |
| *Hysterionica pinifolia* | ≤1 | ≤1 | ≤1 | ≤1 | 2-3 | ≤1 |
| *Baccharis darwinii* | ≤1 | ≤1 | ≤1 | ≤1 | ≤1 | ≤1 |
| *Trichocline sinuata* | ≤1 | ≤1 | ≤1 | ≤1 | ≤1 | ≤1 |
| *Tweedia* sp. | ≤1 | ≤1 | ≤1 | ≤1 | ≤1 | ≤1 |

| References of frequency | |
|---|---|
| ≤1 plots | (light grey) |
| 2-3 plots | (medium grey) |
| 4-5 plots | (dark grey) |

The six sites were: Ungrazed site (UG), Medium grazed site (MG), Heavily grazed site (HG), Ungrazed for a long period then grazed site (LU), Burnt then ungrazed site (BU) and Burnt then grazed site (BG). (Modified from Peter et al. 2013).

On PC2 the burnt sites were separated from the unburnt ones, particularly for the lower frequencies of *Larrea divaricata* and *Prosopidastrum globosum* on unburnt sites. Based on the frequency of each species on each site (Table 2), the frequency of some species of low frequency (shrubs: *Geoffroea decorticans, Larrea cuneifolia, Cyclolepis genistoides, Mulguraea ligustrina* and *Monttea aphylla*; sub-shrubs: *Acantholippia seriphioides* and *Baccharis crispa*; grasses: *Melica rigida, Piptochaethium napostaense* and *Jarava neaei;* forbs: *Baccharis darwinii, Trichocline sinuata* and *Tweedia* sp.) and species of high frequency (shrubs: *Condalia microphylla, Lycium* spp. and *Prosopis alpataco*) did not change with land-use history.

On the one hand, the frequency of some species decreased at higher grazing levels (i.e., *Chuquiraga erinacea, Schinus johnstonii, Senna aphylla, Margyricarpus pinnatus, Baccharis trimera, B. gilliesii, Poa ligularis, Senecio filaginoides* and *S. melanopotamicus*), while on the other hand the frequency of other species increased at higher grazing levels (*Baccharis ulicina, Nassella tenuis* and *Poa lanuginosa*).

On burnt sites the frequency of one shrub decreased (*Larrea divaricata*) whereas the preferred grass, *Pappophorum caespitosum*, increased.

In respect of the species richness of each plant functional type, PC1 explained 70 % of the variability and PC2 25 % (Figure 2). Again, the sites were ordered on the PC1 along a grazing intensity gradient with ungrazed sites having the greater richness of forbs and shrubs. The second axes separated the exclosures from grazing which had been subject to previous burning or grazing (sites BU and LU) with their higher richness of preferred grasses and sub-shrubs.

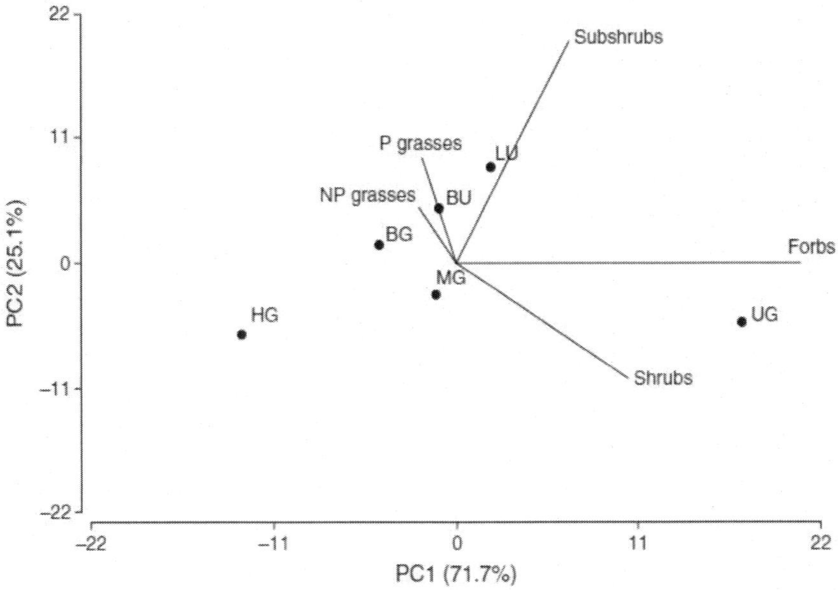


Figure 2. Principal Components Analysis showing the relationships of similarities between sites with respect to the species richness of each plant functional type. The six sites were: Ungrazed site (UG), Medium grazed site (MG), Heavily grazed site (HG), Ungrazed for a long period then grazed site (LU), Burnt then ungrazed site (BU) and Burnt then grazed site (BG). P grasses are preferred grasses and NP grasses are non-preferred grasses.

Modified from Peter et al. 2013.

Species aerial cover and plant functional type cover showed some differences in the ordination of sites. For species cover, PC1 explained 53 % of the variability and PC2 35 % (Figure 3a). Site UG was very different from all other sites with a greater cover of *P. ligularis* and *N. sericeum*. Burnt sites (sites BU and BG) and site LU were defined by a higher cover of *C. erinacea*, while sites MG and HG principally consisted of *L. divaricata.*

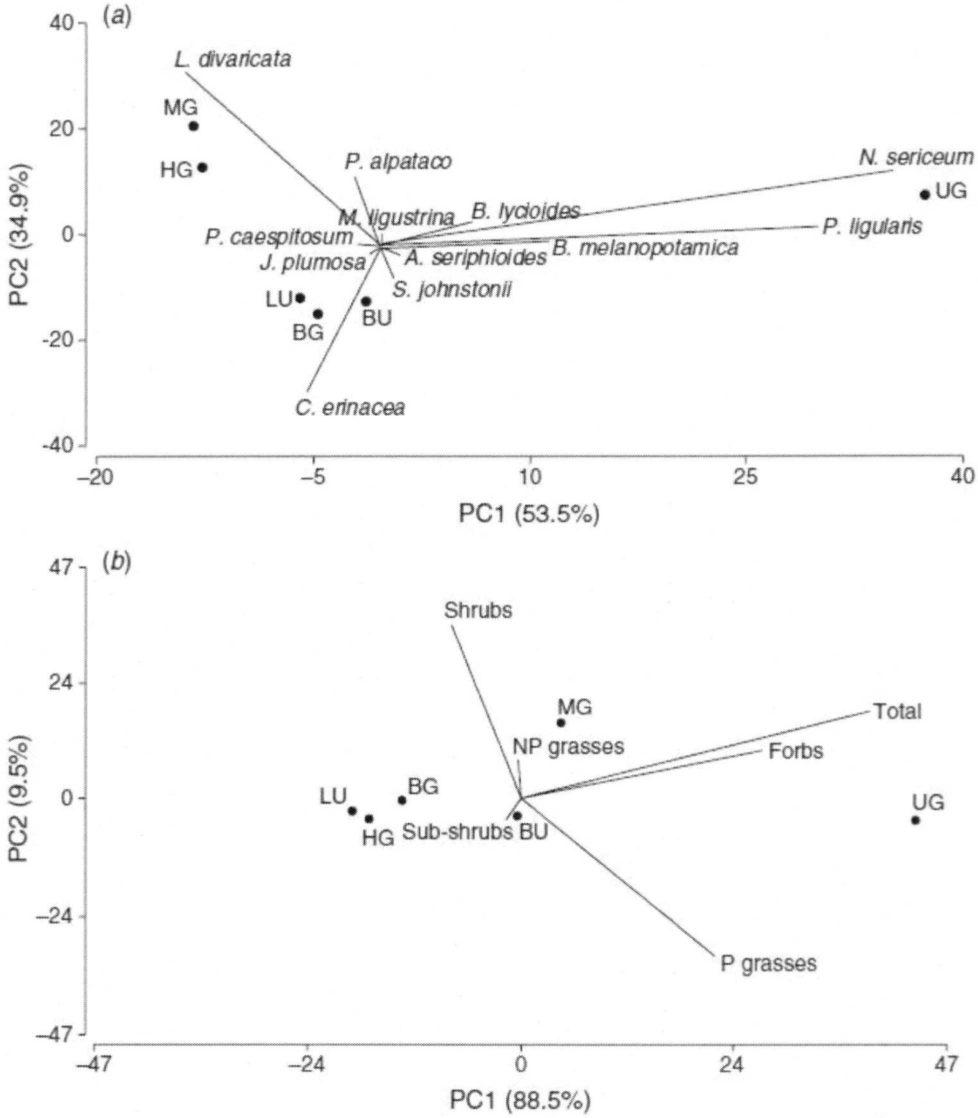


Figure 3. Principal Components Analysis showing the relationships of similarities between sites with respect to the a) specific cover and b) cover of each plant functional type. Only species with a correlation of greater than 0.80 are indicated. The six sites were: Ungrazed site (UG), Medium grazed site (MG), Heavily grazed site (HG), Ungrazed for a long period then grazed site (LU), Burnt then ungrazed site (BU) and Burnt then grazed site (BG).

Modified from Peter et al. 2013.

In the case of cover per plant functional type, PC1 and PC2 explained 88 % and 9 % of the variability respectively (Figure 3b). Site UG showed a higher cover of preferred grasses and forbs, and a higher total cover. However, some grazed sites (sites LU, HG and BG) were characterized by a lower cover of forbs and total cover. Site BU had medium cover for all plant functional types.

The results showed significant differences in the cover of preferred and non-preferred grasses, forbs and total cover between sites (H= 16.98, *P*= 0.004; H= 17.05, *P*= 0.003; H= 18.26, *P*= 0.002; H= 11.94, *P*= 0.03, respectively) and no significant differences were found in the cover of shrubs and sub-shrubs (Figure 4).

Significant differences were principally due to grazing intensity, because the burnt sites did not differ in the cover of any functional group. The site UG had the highest cover of preferred grasses, forbs and total plant species, and the grazed sites had a lower cover of the same elements. For non-preferred grasses, the highest cover was on the BU site and the lowest on the UG site.

## Spatial Distribution per Plant Functional Type

The pattern of spatial distribution changed with land-use history in all plant functional types (Table 3).

**Table 3. Spatial distribution of species and plant functional type at sites with different land use histories**

| **Species** | BG | BU | LU | HG | MG | UG |
|---|---|---|---|---|---|---|
| *Ephedra ochreata* | | | | | | |
| *Geoffroea decorticans* | | | | | | |
| *Monttea aphylla* | | | | | | |
| *Mulguraea ligustrina* | | | | | | |
| *Condalia microphylla* | | | | | | |
| *Brachyclados lycioides* | | | | | | |
| *Prosopidastrum globosum* | | | | | | |
| *Bredemeyera microphylla* | | | | | | |
| *Schinus johnstonii* | | | | | | |
| *Lycium* spp. | | | | | | |
| *Larrea divaricata* | | | | | | |
| *Chuquiraga erinacea* | | | | | | |
| *Prosopis alpataco* | | | | | | |
| **Total Shrubs** | | | | | | |
| *Acantholippia seriphioides* | | | | | | |
| *Baccharis crispa* | | | | | | |
| *Ephedra triandra* | | | | | | |
| *Grindelia chiloensis* | | | | | | |
| *Margyricarpus pinnatus* | | | | | | |
| *Sphaeralcea mendocina* | | | | | | |
| *Baccharis ulicina* | | | | | | |
| *Senna aphylla* | | | | | | |
| **Total Subshrubs** | | | | | | |
| *Melica rigida* | | | | | | |
| *Pappostipa speciosa* | | | | | | |

**Table 3. (Continued)**

| **Species** | BG | BU | LU | HG | MG | UG |
|---|---|---|---|---|---|---|
| **Total Non palatable grasses** | [random] | [random] | [random] | [random] | [aggregated] | [random] |
| *Jarava neaei* | [absent] | [absent] | [random] | [absent] | [absent] | [absent] |
| *Jarava plumosa* | [random] | [random] | [random] | [absent] | [absent] | [absent] |
| *Nassella tenuis* | [absent] | [absent] | [random] | [random] | [absent] | [absent] |
| *Piptochaetium napostaense* | [absent] | [random] | [random] | [absent] | [absent] | [absent] |
| *Pappophorum caespitosum* | [random] | [random] | [random] | [random] | [random] | [absent] |
| *Poa ligularis* | [aggregated] | [random] | [aggregated] | [random] | [random] | [aggregated] |
| *Nasella longiglumis* | [aggregated] | [aggregated] | [absent] | [random] | [random] | [random] |
| *Poa lanuginosa* | [random] | [random] | [random] | [aggregated] | [aggregated] | [random] |
| **Total Palatable grasses** | [aggregated] | [absent] | [aggregated] | [aggregated] | [aggregated] | [aggregated] |
| *Baccharis darwinii* | [random] | [absent] | [absent] | [absent] | [absent] | [random] |
| *Tweedia* sp. | [absent] | [random] | [absent] | [absent] | [absent] | [absent] |
| *Hysterionica pinifolia* | [absent] | [random] | [absent] | [absent] | [absent] | [random] |
| *Senecio melanopotamicum* | [absent] | [absent] | [absent] | [absent] | [random] | [random] |
| *Trichocline sinuata* | [absent] | [absent] | [random] | [absent] | [absent] | [absent] |
| *Boopis anthemoides* | [absent] | [random] | [random] | [absent] | [random] | [regular] |
| *Senecio filaginoides* | [random] | [random] | [random] | [absent] | [random] | [random] |
| **Species** | BG | BU | LU | HG | MG | UG |
| *Noticastrum sericeum* | [random] | [random] | [absent] | [absent] | [random] | [aggregated] |
| *Baccharis melanopotamica* | [random] | [aggregated] | [random] | [absent] | [random] | [aggregated] |
| *Baccharis gilliesii* | [random] | [aggregated] | [random] | [absent] | [aggregated] | [aggregated] |
| **Total Forbs** | [random] | [aggregated] | [regular] | [absent] | [aggregated] | [aggregated] |

| References of distribution pattern | |
|---|---|
| absent | |
| random | |
| regular | |
| aggregated | |

Data were obtained from the species cover and cover of each plant functional type. The six sites were: Ungrazed site (UG), Medium grazed site (MG), Heavily grazed site (HG), Ungrazed for a long period then grazed site (LU), Burnt then ungrazed site (BU) and Burnt then grazed site (BG). (Modified from Peter et al. 2013).

For shrubs the pattern changed from random in the ungrazed sites to aggregate in the grazed sites. The pattern of sub-shrubs was aggregated in HG and MG sites and random in the others. The forbs group showed the greatest changes; it was absent in site HG, regular in site LU, random in site BG and aggregated in the other sites. Preferred grasses presented an aggregated pattern except in site BU, where it was random. Amongst non-preferred grasses, the pattern was aggregated in sites UG and MG, random in site HG and absent in the other sites. This pattern was driven primarily by one species - *P. speciosa* - which was the principal non-preferred grass.

In general, the ungrazed and burnt areas changed in pattern from aggregated to random, probably due to the removal of niches and the homogenization of resources and propagules.

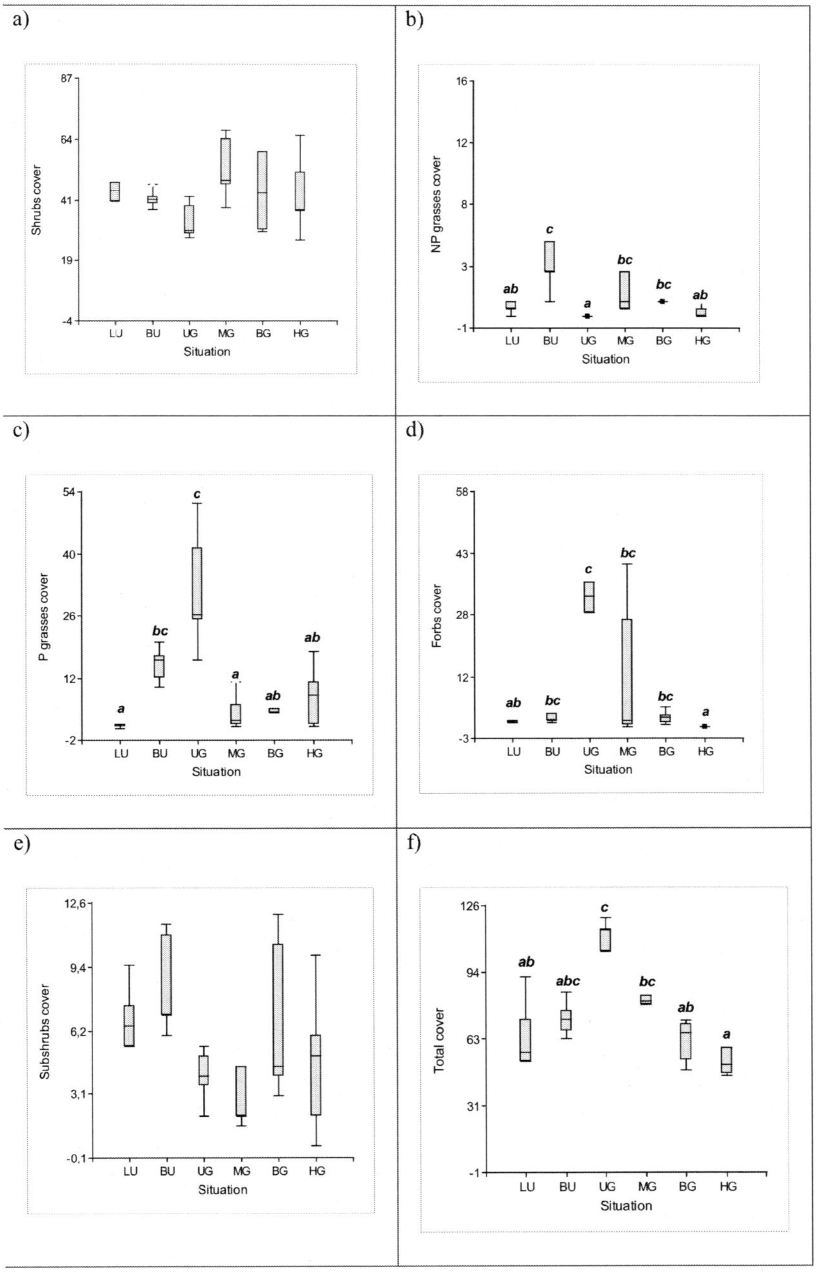


Figure 4. Cover ($m^2$) of each plant functional type. a) shrubs; b) non-preferred grasses; c) preferred grasses; d) forbs; e) sub-shrubs; and f) total cover. The solid line within each box represents the median. Horizontal hyphens at the top and bottom of vertical lines represent the maximum and minimum values, respectively. The lower Q1 and the upper Q3 quartiles are represented as the bottom and top horizontal lines of each box. Vertical bars indicate the distance between the minimum and maximum values. Different letters indicate significant differences ($P< 0.05$) in the cover of each plant functional type. The absence of letters indicates no detection of significant differences within a plant functional group. The six sites were: Ungrazed site (UG), Medium grazed site (MG), Heavily grazed site (HG), Ungrazed for a long period then grazed site (LU), Burnt then ungrazed site (BU) and Burnt then grazed site (BG).

Modified from Peter et. al. 2013.

## Stocked Carbon in Soil

In respect to the soil variables, PC1 and PC2 explained 60.5 % of the variability (Figure 5). Axis 1 was influenced principally by the SOC associated with the particles of different sizes. Sites were ordered in a disturbance gradient on PC1, with the HG site associated positively with the fraction SOC and UG site on the opposite side; burnt sites in mean positions and MG and LU distributed along the axis (Figure 5).

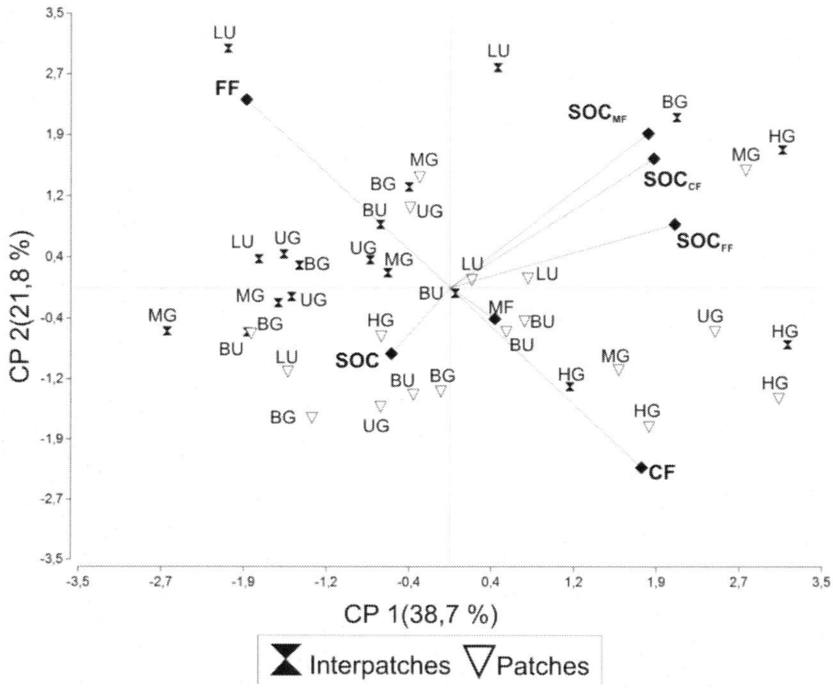


Figure 5. Principal Components Analysis showing the relationships of similarities between sites with respect to the means values of distribution of soil particle sizes (g): coarse fraction (CF), medium fraction (MF) and fine fraction (FF); means values of organic fractions of soil (t $ha^{-1}$): organic carbon in coarse fraction ($SOC_{CF}$), organic carbon in medium fraction ($SOC_{MF}$) and organic carbon in fine fraction ($SOC_{FF}$) and means values of soil organic carbon SOC. The six sites were ungrazed for a long period then grazed site (LU), ungrazed site (UG), burnt then ungrazed site (BU), medium-grazed site (MG), burnt then grazed site (BG) and heavily-grazed site (HG).

In this analysis, mostly of patches and interpatches are separated mainly by the size of particles, with patches associated with the coarse particles (greater than 105 μm) and greater particulate SOC, and interpatches related with fine particles (lower than 53 μm) (Figure 5). At this point is remarkable that UG and HG, the sites in the extremes of the land use gradient, did not show differences between patches and interpatches, with the UG samples tended to the second quadrant (fine particles) and the HG samples to the fourth quadrant (coarse particles). SOC did not show significant differences among situations neither in patches nor in interpatches.

Respect to soil size fractions in interpatches of sites with different land use histories showed highly significant differences in the weight of CF and FF, and significant differences for MF (Table 4). HG presented the greatest weight of CF and the lowest of FF than in rest of the sites; whereas the opposite variation (lower CF and higher FF) was observed in sites with prolonged exclosures (LU) compared with HG, BU and BG. Sites that had been exposed to events of moderated disturbance (BU, MG and BG) did not show marked differences in the amount of particles of different sizes in the interpatches, but BG presented greater amount of MF than MG (Table 4).

In the patches, only the particles of the MF presented differences among sites in relation with different uses, the burnt sites and LU has greater weight of MF than HG (Table 5).

The BD showed significant differences in interpatches and patches (Tables 4 and 5). In the interpatches, LU showed the highest value of BD and UG had the lowest value compared to other sites. In the patches, BD was higher in BG that in BU, UG and MG.

In the analyses of SOC associated with the different sizes of particles, the interpatches showed differences between land uses only in the $SOC_{FF}$. The most intense grazed site (HG) had greater amount of organic carbon in comparison with LU, UG, BU and MG (Table 4). Patches had differences between land uses only in the $SOC_{CF}$, with the greatest amount present in MG than in rest of the sites, except with HG (Table 5). The other sites show no significant differences between them (Table 5).

# DISCUSSION

## Vegetation Attributes

The results revealed that land-use history influenced all the vegetation attributes analyzed (frequency, richness, cover and spatial pattern) in all functional groups.

**Table 4. Soil properties and stocked organic carbon in soil in interpatches**

| Interpatches | LU | BU | UG | MG | BG | HG | p-valor |
|---|---|---|---|---|---|---|---|
| CF (g soil) | 17.5±1.3$^{a}$ | 21.8±1.0$^{b}$ | 21.5±0.7$^{b}$ | 21.7±0.6$^{b}$ | 19.0±0.8$^{ab}$ | 28.1±1.6$^{c}$ | 0.0004 |
| MF (g soil) | 9.3±1.0$^{a}$ | 10.6±0.1$^{ab}$ | 9.4±0.9$^{a}$ | 8.9±0.3$^{a}$ | 12.1±0.7$^{b}$ | 10.6±0.3$^{ab}$ | 0.035 |
| FF (g soil) | 21.7±2.3$^{c}$ | 16.1±1.2$^{b}$ | 17.6±0.8$^{bc}$ | 17.9±0.6$^{bc}$ | 17.2±0.7$^{b}$ | 10.6±1.6$^{a}$ | 0.0026 |
| BD (t $m^{-3}$) | 1.48±0.01$^{c}$ | 1.35±0.03$^{b}$ | 1.23±0.02$^{a}$ | 1.37±0.03$^{b}$ | 1.34±0.02$^{b}$ | 1.39±0.04$^{b}$ | 0.0009 |
| SOC (t CO $ha^{-1}$) | 4.7±0.7 | 4.7±0.8 | 3.7±0.9 | 5.7±0.2 | 5.7±0.3 | 4.1±0.1 | 0.1817 |
| $SOC_{CF}$ (t CO $ha^{-1}$) | 5.4±0.7 | 4.1±0.8 | 3.3±0.6 | 3.1±0.9 | 6.6±1.8 | 5.9±1.0 | 0.1885 |
| $SOC_{MF}$ (t CO $ha^{-1}$) | 5.0±1.3 | 3.6±0.7 | 3.5±0.4 | 2.5±0.5 | 4.1±0.5 | 5.4±1.1 | 0.215 |
| $SOC_{FF}$ (t CO $ha^{-1}$) | 2.5±0.2$^{a}$ | 2.4±0.1$^{a}$ | 2.4±0.1$^{a}$ | 2.4±0.1$^{a}$ | 2.7±0.2$^{ab}$ | 3.02±0.1$^{b}$ | 0.0309 |

Mean values (± SE) of soil particle sizes (g) in interpatches: coarse fraction (CF), medium fraction (MF) and fine fraction (FF); mean values of bulk density (BD); mean values of organic fractions of soil (t ha-1): organic carbon in coarse fraction ($SOC_{CF}$), organic carbon in medium fraction ($SOC_{MF}$) and organic carbon in fine fraction ($SOC_{FF}$); and mean values of soil organic carbon SOC. The six sites were: Ungrazed site (UG), Medium grazed site (MG), Heavily grazed site (HG), Ungrazed for a long period then grazed site (LU), Burnt then ungrazed site (BU) and Burnt then grazed site (BG). Different letters indicate significant differences among means. The results are shown without transforming.

Species frequency was especially influenced by grazing intensity, with fire as a lesser source of variability. Neither dominant nor rare species changed frequency. The frequency of the dominant species of the community, e.g., *Condalia microphylla* and *Prosopis alpataco*, did not vary significantly, indicating their ability to tolerate all levels of disturbance. Amongst the species that declined in frequency under higher grazing intensities, the only preferred species was *Poa ligularis*, so its decrease was probably due to the loss of individuals as a consequence of defoliation. In the case of non-preferred species (*Chuquiraga erinacea, Schinus johnstonii, Senna aphylla, Margyricarpus pinnatus, Baccharis trimera, B. gilliesii, Senecio filaginoides* and *S. melanopotamicus*), the diminished frequency could be due to the lack of optimal sites for recruitment as a result of trampling. At higher grazing pressures, the frequency of the preferred grasses, *Poa lanuginosa* and *Nassella tenuis*, increased. *Poa lanuginosa* is a rhizomatous grass which has been reported as a colonizer of bare ground due to its vegetative reproduction (Busso and Bonvissuto 2009). It is possible that individuals that have survived grazing form new tillers from their rhizomes in sites without cover. Rostagno et al. (2006) reported that both these species are able to establish readily in mounds after fires, especially *N. tenuis*, which they found dominated the shrub interspaces which are characterized by a soil with low nitrogen and phosphorus contents. This did not occur in this study. Rather, these species decreased at the burnt sites. The lower frequency of the shrub, *Larrea divaricata*, in burnt sites might be due to its high mortality (approx. 60 %) after wildfires (Bóo et al. 1997). In contrast, the frequency of the preferred grass, *Pappophorum caespitosum*, increased in burnt sites as this species prefers open sites (Villagra 2000), and fire diminishes aerial cover and provides sites for recruitment.

**Table 5. Soil properties and stocked organic carbon in soil in patches**

| Patches | LU | BU | UG | MG | BG | HG | p-valor |
|---|---|---|---|---|---|---|---|
| CF (g soil) | 22.1±0.06 | 25.5±0.6 | 25.3±2.5 | 24.4±2.0 | 22.2±1.7 | 30.7±2.4 | 0.0581 |
| MF (g soil) | 11.8±0.5$^{bc}$ | 10.9±0.7$^{bc}$ | 10.8±0.4$^{ab}$ | 10.3±1.3$^{ab}$ | 13.0±0.5$^{c}$ | 8.3±0.3$^{a}$ | 0.0122 |
| FF (g soil) | 14.8±0.9 | 12.1±0.2 | 13.1±2.8 | 13.8±3.3 | 13.5±1.5 | 9.6±2.5 | 0.6174 |
| BD (t $m^{-3}$) | 1.27±0.01$^{cd}$ | 1.16±0.02$^{ab}$ | 1.14±0.03$^{a}$ | 1.21±0.04$^{abc}$ | 1.30±0.03$^{d}$ | 1.24±0.04$^{bcd}$ | 0.0113 |
| | | | | | | | |
| SOC (t CO $ha^{-1}$) | 1.8±0.5 | 1.5±0.07 | 1.5±0.3 | 1.4±0.3 | 1.7±0.4 | 1.6±0.2 | 0.9015 |
| $SOC_{CF}$ (t CO $ha^{-1}$) | 4.2±0.7$^{ab}$ | 4.5±0.6$^{ab}$ | 4.8±1.2$^{ab}$ | 7.2±0.2$^{c}$ | 3.4±0.5$^{a}$ | 5.9±0.5$^{bc}$ | 0.024 |
| $SOC_{MF}$ (t CO $ha^{-1}$) | 1.2±0.1 | 1.2±0.2 | 1.3±0.4 | 1.4±0.24 | 0.71±0.2 | 1.3±0.3 | 0.4803 |
| $SOC_{FF}$ (t CO $ha^{-1}$) | 16.4±2.0 | 13.4±1.1 | 14.9±0.1 | 17.5±2.6 | 11.2±0.3 | 14.5±3.2 | 0.3249 |

Mean values (± SE) of soil particle sizes (g) in vegetation patches: coarse fraction (CF), medium fraction (MF) and fine fraction (FF); mean values of bulk density (BD); mean values of organic fractions of soil (t ha-1): organic carbon in coarse fraction ($SOC_{CF}$), organic carbon in medium fraction ($SOC_{MF}$) and organic carbon in fine fraction ($SOC_{FF}$); and mean values of soil organic carbon SOC. The six sites were: Ungrazed site (UG), Medium grazed site (MG), Heavily grazed site (HG), Ungrazed for a long period then grazed site (LU), Burnt then ungrazed site (BU) and Burnt then grazed site (BG). Different letters indicate significant differences among means. The results are shown without transforming.

Land-use histories affected the species richness of each plant functional type in a similar way. Ungrazed areas were characterized by greater richness of forbs and shrubs, but the highest richness of grasses (preferred and non-preferred) was in the burnt areas, regardless of whether they were being grazed or not. Peláez et al. (2010) suggested that this response is explained, at least in part, by reduced competition of shrubs for light, water and/or nutrients after fire. On the other hand, this might be due to the provision of sites for recruitment which have inherently better growing conditions, enabling a large number of species to be recruited, thus reducing the effects of grazing. Previous work in semi-arid woodlands showed that fires contribute to grasses re-sprouting and limit shrub growth, altering community composition from a shrubland to a grassland (Hodgkinson and Harrington 1985; West and Hassan 1985; Wambolt and Payn 1986; Pfeiffer and Steuter 1994; Bóo et al. 1996; Rostagno et al. 2006; Ravi et al. 2007; Peláez et al. 2010).

In general, the results are according to our first prediction because they suggest that grazing diminishes the cover of grasses, forbs and total cover. Ordination of sites in respect to specific cover supports the idea that shrubs of dense canopies, such as *C. erinacea*, are replaced by open canopied shrubs, such as *L. divaricata*, under grazing (Bisigato et al. 2005; Fernández et al. 2009) because the latter favour the passage of cattle. Besides, *L. divaricata* has a greater recruitment in bare soil under grazing (Bisigato and Bertiller 1999) and is more affected by fire (Bóo et al. 1997). Thus, differences found in the ordination of grazed and burnt sites with reference to species cover and cover of functional groups could be explained by the replacement of species of the same functional group, for example *C. erinacea* by *L. divaricata*. Contrary to our second prediction, no differences were found in the cover of shrubs and sub-shrubs between the different sites. Accordingly, there is no evidence of shrub encroachment in this system. This conclusion coincides with that of Cipriotti and Aguiar (2005, 2012) and Cesa and Paruelo (2011) and is partially coincident with that of Rostagno et al. (2006) who concluded that removal of livestock favours the recovery of grasses but does not change shrub dominance. Bran et al. (2007) found differences in the shrub cover between unburnt and burnt places 21 months after the fire. Moreover, Peláez et al. (2010) showed that the cover of some shrub species of Monte region recovered five years after a fire, which could explain the absence of significant differences between the sites in this study. On the other hand, the burnt ungrazed site had the highest cover of non-preferred grasses, and had medium values for the other functional groups. In this context, Franzese and Ghermandi (2012) mentioned that burning positively affects the emergence of P. *speciosa,* the most important non-preferred grass in the area. This is in agreement with our hypothesis that fire attenuates the grazing effects on vegetation cover because with fire there is a non-selective removal of biomass and because fire promotes wind erosion of soil and redistribution of sediments and propagules. Besides, no difference was found in forbs, preferred grasses and total cover between the burnt site, excluded from grazing, and the ungrazed area, which might be because fire accelerates the recovery processes of grasses due to reduced competition from shrubs (Peláez et al. 2010), and the aerial cover of vegetation attained similar values to those of the sites excluded from grazing.

## Spatial Distribution

Consistent with our third prediction, the spatial distribution pattern of some species and of all plant functional types was found to change in relation to land use. As mentioned in the Results section, the vegetation pattern changed from aggregated under ungrazed conditions to random in ungrazed and burnt areas. As in other arid shrubby systems of the world, fire tends to break the patch and interpatch pattern and to accelerate recovery of vegetation in areas of bare soil (Ravi et al. 2009; Whicker et al. 2002). This is probably due to the destruction of niches which then allows the random recruitment of new plants.

## Stocked Carbon in Soil

Land use history affects mainly the distribution of particles of the solid matrix of the soil. We focused principally in interpatches because these are the micro-ambient expected to be revegetated. In interpatches, burnt and grazed site (BG) has weight of medium fraction greater than medium grazed and ungrazed sites. The difference in disturbance among these sites was the occurrence of both, fire and grazing. Sites affected only by fire (burnt and ungrazed site) did not show differences with exclosure. Particles of medium size are associated with the useful carbon (Kapkiyai 1994; Galantini et al. 2008; Duval et al. 2013; Toledo et al. 2013). Then, the occurrence of both disturbances could be useful at the moment of apply management tools with the finality of increasing the useful carbon in the interpatches. In highly grazed sites, coarse fraction augmented. The interpatches in heavily grazed sites have more proportion of bare soil (Gaitán et al. 2009), and the particles of minor sizes (FF and MF) are more susceptible to move when the superficial cover is scarce (Buschiazzo and Aimar 2003). In similar situations, Peter et al. (2011) found that with the exclusion of the area, the high of the mounds associated to the vegetation patches diminished, probably by the discontinuation of movements of material from the interpatches to the patches (Ravi et al. 2009).

In the soil of patches, the only difference was in the weight of medium size fraction. Burnt ungrazed site have similar proportion of particles than ungrazed site, but burnt and grazed site is different of exclosure. Thus, the fire would produce the homogenization in the distribution of particles through wind erosion, while grazing would increase the erosion process. Soil loss occurs, probably, because the medium fraction is one of the more susceptible to distribution from the interpatches to the patches (Buschiazzo and Aimar 2003). Fire could disrupt the pattern of patches and interpatches, but only when it is follow by exclousures.

In the interpatches, the ungrazed site has a lower BD. This is consistent with other studies where livestock removal leads to a decrease in soil compaction and to increases in infiltration rate, soil aggregate stability and soil nutrient concentration (Allington and Valone 2010). Furthermore, the site with prolonged exclusion and subsequent reintroduction of grazing livestock (LU) has a high BD. This might be due the sensitivity of the system to the reintroduction of grazing after a long time of exclosure, thus leading to an increase in land degradation.

Our results are according with a previous study in the region (Gaitán et al. 2009), we did not find differences in the total stocked soil carbon. This supports the hypothesis that SOC is not a reliable indicator of soil fertility associated with different land uses (Galantini and Rosell 2006). Because of that, these authors recommend the evaluation of the soil organic carbon fractions in spite of the total organic carbon.

The study of the soil organic fraction in interpatches showed that the most disturbed site (HG) presented greatest $SOC_{FF}$ than all the other situations (except BG). But, this stocked organic carbon is not available to the vegetation because it is associated with the mineral fraction (finest fraction) of the soil (Post and Kwon 2000; Galantini and Rosell 2006; Kutsch et al. 2009). In patches, the medium intensity of grazing increased the carbon content associated with the CF in relation with the exclosure. In opposition with our results for soil size particle fraction, the content of carbon in the soil was not augment by fire. Burnt and grazed sites have lowest carbon content, and it is not different of other situations (nor burnt, nor excluded). These contradict our fourth prediction of greater SOC in exclosures and lower SOC in burnt sites. Our results suggest that a moderate disturbance (fire followed by grazing or medium grazing intensity) enlarged the SOC and the amount of rapidly available carbon in the soil. Several authors have found that grazing decreases the concentration of soil nutrients, such as organic matter and total nitrogen (Rostagno and Beeskow 2000; Chartier et al. 2011). Depending on the system degradation state, the removals of livestock grazing enhance the soil properties. In the absence of trampling by livestock there is a reduction of erosion and an increase in the accumulation of nutrients in the soil surface. However, these changes can take decades to occur (Castellano and Valone 2007; Allington and Valone 2014).

## Conclusion

The results of this study support the view that this semi-arid rangeland type is resilient to fire (Bóo et al. 1997) but not so to continuous grazing (Rostagno et al. 2006). In spite of not finding shrub encroachment, a positive response of the grasses to fire and exclusion was found, and previous studies (Bóo et al. 1997; Rostagno et al. 2006; Tizón et al. 2010) have suggested that repeated fires may be necessary to suppress woody plants and favors grass species. On the other hand, an intermediate level of disturbance from livestock grazing can promote a more heterogeneous landscape, providing greater opportunities for a variety of plant species (McNaughton 1983); that is, moderate grazing may sustain a greater diversity of native plant species and a greater amount of organic carbon in the soil. There are doubts about the efficacy of short-term removal of cattle for increasing plant diversity at a local scale and also about moderate grazing practices that may result in increased community heterogeneity (Loeser et al. 2007).

According to Fuhlendorf and Engle (2001), landscape management which involves a heterogeneous patchwork of disturbance, such as grazing or fire, followed by periods of rest, is recommended for increasing forage production and content of organic carbon in the soil. The low grazing intensity combined with fires allow to the system to recover more rapidly (Ravi and D'Odorico 2009).

## Acknowledgment

We thank the staff and owners of the sites. This work was supported with grants from ANPCyT and Universidad Nacional de Río Negro (PICT-O 198/10 and PIUNRN 40-C-164). GP and MAL are members of CONICET.

## References

Aguiar, M. R., and Sala, O. A. (1999).Patch structure, dynamics and implications for the functioning of arid ecosystems. *Trends in Ecology and Evolution,* 14, 273–277.

Allington, G. R. H., and Valone, T. J. (2010). Reversal of desertification: the role of physical and chemical soil properties. *Journal of Arid Environments,* 74, 973-977.

Allington, G. R. H., and Valone, T. J. (2011). Long-term livestock exclusion in an arid grassland alters vegetation and soil. *Rangeland Ecology and Management*, 64, 424–428.

Allington, G. R. H., and Valone, T. J. (2014). Islands of fertility: a byproduct of grazing? *Ecosystems,* 17(1), 127-141.

Arturi, M., Torres Robles, S. S., Herrera Santángelo, R., UNLP, FCAyF, and CFI. (2011). Proceso de ordenación territorial de los recursos naturales renovables del Monte y Espinal de la provincia de Río Negro. Ciudad Autónoma de Buenos Aires: Consejo Federal de Inversiones (CFI). http://cfired.org.ar/Default.aspx?mod=databasesandtype=2

Balzarini, M. G., Gonzalez, L., Tablada, M., Casanoves, F., Di Rienzo, J. A., and Robledo, C. W. (2008). Infostat. Manual del Usuario. Córdoba: Editorial Brujas.

Braun-Blanquet, J. (1972). Plant sociology: the study of plant communities. New York:Hafner.

Buschiazzo, D. E., and Aimar, S. B. (2003). 2. Erosión eólica: procesos y predicción. In: A. D. Gollberg, and A. G. Kin (Eds.), Viento, suelo y plantas. Buenos Aires: Ediciones INTA.

Busso, C. A., and Bonvissuto, G. L. (2009).Structure of vegetation patches in north-western Patagonia, Argentina. *Biodiversity and Conservation*, 18, 3017–3041.

Busso, C. A., Bonvissuto, G. L., and Torres, Y. A. (2012).Seedling recruitment and survival of two desert grasses in the Monte of Argentina. *Land Degradation and Development*, 23, 116–129.

Cabrera, A. L. (1971). Fitogeografía de la República Argentina. *Boletín de la Sociedad Argentina de Botánica,* 16, 1-42.

Castellano, M. J., and Valone, T. J. (2007). Livestock, soil compaction and water infiltration rate: evaluating a potential desertification recovery mechanism. *Journal of Arid Environments*, 71, 97-108.

Cesa, A., andParuelo, J. M. (2011). Changes in vegetation structure induced by domestic grazing in Patagonia (Southern Argentina). *Journal of Arid Environments,* 75, 1129-1135.

Chartier, M. P., Rostagno, C. M., and Pazos, G. E. (2011). Effects of soil degradation on infiltration rates in grazed semiarid rangelands of northeastern Patagonia, Argentina. *Journal of Arid Environments*, 75, 656-661.

Cingolani, A. M., Noy-Meir, I., and Díaz, S. (2005). Grazing effects on rangeland diversity: a synthesis of contemporary models. *Ecological Applications,* 15, 757–773.

Cingolani, A. M., Noy-Meir, I., Renison, D. D., and Cabido, M. (2008). La ganadería extensiva, es compatible con la conservación de la biodiversidad y de los suelos? *Ecología Austral.*, 18, 253–271.

Cipriotti, P. A., and Aguiar, M. R. (2005).Effects of grazing on patch structure in a semi-arid two-phase vegetation mosaic. *Journal of Vegetation Science,* 16, 57–66.

Cipriotti, P. A., and Aguiar, M. R. (2012). Direct and indirect effects of grazing constrain shrub encroachment in semi-arid Patagonian steppes. *Applied Vegetation Science,* 15, 35-47.

de la Cruz Rot, M. (2008). Capítulo 3: Métodos para analizar datos puntuales. In: Introducción al Análisis Espacial de Datos en Ecología y Ciencias Ambientales: Métodos y Aplicaciones. InF. T. Maestre, A. Escudero, and A. Bonet (Eds.), Introduction to the Spatial Analysis of Information in Ecology and Environmental Sciences: Methods and Applications. Madrid: Dykinson.

Di Rienzo, J. A., Casanoves, F., Balzarini, M. G., Gonzalez, L., Tablada, M., and Robledo, C. W. (2011). InfoStat versión 2011. Córdoba: Grupo InfoStat, FCA, Universidad Nacional de Córdoba.

Digby, P. G. N., and Kempton, R. A. (1987). Multivariate analysis of ecological communities. London: Chapman and Hall.

D'Odorico, P., Caylor, K., Okin, G. S., and Scanlon, T. M. (2007). On soil moisture–vegetation feedbacks and their possible effects on the dynamics of dryland ecosystems. *Journal of Geophysical Research: Biogeosciences,* (2005–2012), 112(G4).

Duval, M. E., Galantini J. A., Iglesias, J. O., Canelo S., Martínez J. M., and Wall, L. (2013). Analysis of organic fractions as indicators of soil quality under natural and cultived systems. *Journal Soil and Tillage Research*, 131, 11-19.

FAO. (2002). Captura de carbono en los suelos para un mejor manejo de la tierra. Basado en el trabajo de Michel Robert, Institut National de Recherche Agronomique. París: FAO.

Fernández, O. A., Gil, M. E., and Distel, R. A. (2009). The challenge of rangeland degradation in a temperate semi-arid region of Argentina: the Caldenal. *Land Degradation and Development*, 20, 431-440.

Fleischner, T. L. (1994). Ecological costs of livestock grazing in western North America. *Conservation Biology*, 8, 629-644.

Franzese, J., and Ghermandi, L. (2012).Effect of fire on recruitment of two dominant perennial grasses with different palatability from semi-arid grasslands of NW Patagonia (Argentina). *Plant Ecology*, 213, 471-481.

Fuhlendorf, S. D., and Engle, D. M. (2001).Restoring heterogeneity on rangelands: ecosystem management based on evolutionary grazing patterns. *Bio. Science*, 51, 625-632.

Gaitán, J. J., Bran, D. E., and Murray, F. (2007) Efecto de la severidad de quemado sobre la concentración de carbono orgánico en montículos e intermontículos en el Monte Austral. *Ciencia del Suelo*, 25, 195-199.

Gaitán, J. J., López, C. R., and Bran, D. E. (2009). Efectos del pastoreo sobre el suelo y la vegetación en la Estepa Patagónica. *Ciencia del Suelo*, 25, 53-63.

Galantini J. A., Iglesias J. O., Landriscini, M., Suñer, L., and Minoldo, G. (2008). Calidad y dinámica de las fracciones orgánicas en sistemas naturales y cultivados. In: J. A. Galantini (Ed.), Estudio de las fracciones orgánicas en suelos de la Argentina. Bahía Blanca: EdiUNS.

Galantini, J. A., and Rosell, R. (2006). Chapter 17: Effects of Environmental and Management Practices on the Potential for Climatic Change Mitigation in the Pampas of Argentina. In: R. Lal, C. C. Cerri, M. Bernoux, J. Etchevers, and C. E. P. Cerri (Eds.), Carbon Sequestration in Soils of Latin America. Nueva York: The Haworth Press Inc.

Godagnone, R. E., and Bran, D. E. (Eds). (2009). Inventario Integrado de los Recursos Naturales de la Provincia de Río Negro. Buenos Aires: INTA.

Hodgkinson, K. C., and Harrington, G. N. (1985).The case for prescribed burning to control shrubs in eastern semi-arid woodlands. *The Australian Rangeland Journal,* 7, 64–74.

Kapkiyai, J. J., Karanja, N. K., Qureshi, J. N., Smithson, P. C., and Woomer, P. L. (1999). Soil organic matter and nutrient dynamics in a Kenyan nitisol under long-term fertilizer and organic input management. *Journal Soil Biology and Biochemistry,* 31, 1773-1782.

Kröpfl, A. I., Cecchi, G. A., Villasuso, N. M., and Distel, R. A. (2013). Degradation and recovery processes in semi-arid patchy rangelands of Northern Patagonia, Argentina. *Land Degradation and Development*, 24, 393-399.

Kutsch, W. L., Bahn, M. and Heinemeyer, A. (2009). Soil Carbon Dynamics: An Integrated Methodology. New York: Cambridge University Press.

Loeser, M. R. R., Sisk, T. D., and Crews, T. E. (2007). Impact of grazing intensity during drought in an Arizona grassland. *Conservation Biology,* 21, 87–97.

Loydi, A., Zalba, S. M., and Distel, R. A. (2012). Vegetation change in response to grazing exclusion in montane grasslands, Argentina. *Plant Ecology and Evolution,* 145, 313–322.

Mateucci, S. D., and Colma, A. (1982). Metodología para el estudio de la vegetación. Monografía Nº 22, Serie Biología. Washington D.C.: Secretaría General de la OEA.

McNaughton, S. J. (1983). Serengeti grassland ecology: the role of composite environmental factors and contingency in community organization. *Ecological Monographs*, 53, 291–320.

Milchunas, D. G., and Lauenroth, W. (1993). Quantitative effects of grazing on vegetation and soils over a global range of environments. *Ecological Monographs,* 6, 327–366.

Mueller-Dombois D., and Ellenberg, H. (1974).Aims and methods of vegetation ecology. New York: Wiley.

Oesterheld, M., Loreti, J., Semmartin, M. and Paruelo, J. M. (1999). Grazing, fire and climate effects on primary productivity of grasslands and savannas. In: Walker, L. (Eds), Ecosystems of Disturbed Ground. Oxford: Elsevier Science.Olff, H., and Ritchie, M. E. (1998). Effects of herbivores on grassland plant diversity. *Science,* 13, 261–265.

Parizek, B., Rostagno, C. M., and Sottini, R. (2002). Soil erosion as affected by shrub encroachment in North-eastern Patagonia. *Journal of Range Management,* 55, 43–48.

Peláez, D. V., Bóo, R. M., Elia, O. R., and Mayor, M. D. (2003). Effect of fire on growth of three perennial grasses from central semi-arid Argentina. *Journal of Arid Environments,* 55, 657- 673.

Peláez, D. V., Giorgetti, H. D., Montenegro, O. A., Elía, O. R., Rodríguez, G. D., Bóo, R. M., and Mayor, M. D. (2010). Vegetation response to a controlled fire in the phytogeographical province of the Monte, Argentina. *Phyton,* 79, 169–176.

Peter, G., Funk, F. A., and Torres Robles, S. S. (2013). Responses of vegetation to different land-use histories involving grazing and fire in the North-east Patagonian Monte, Argentina. *The Rangeland Journal*, 35, 273-283.

Peter, G., Funk, F. A., Loydi, A., Leder, C. V. and Casalini, A. I. (2011). Montículos fitogénicos bajo diferentes presiones de pastoreo en el Monte rionegrino. *Bioscriba,* 4, 44-46.

Peter, G., Funk, F. A., Loydi, A., Casalini, A. I., and Leder, C. V. (2012). Variación de la composición específica y cobertura del pastizal bajo diferentes presiones de pastoreo en el Monte rionegrino. *Phyton,* 81, 4-11.

Pfeiffer, K. E., and Steuter, A. A. (1994).Preliminary response of sand-hills prairie to fire and bison grazing. *Journal of Range Management*, 47, 395–397.

Post, W. M. and Kwon, K. C. (2000). Soil carbon sequestration and land-use change: processes and potential. *Journal Global Change Biology*, 6, 317–327.

Puigdefábregas, J. (1998). Ecological impacts of global change on drylands and their implications for desertification. *Land Degradation and Development,* 9, 393-406.

Ravi, S., and D'Odorico, P. (2009). Post-fire resource redistribution and fertility island dynamics in shrub encroached desert grasslands: a modeling approach. *Landscape Ecology*, 24, 325-335.

Ravi, S., D'Odorico, P., Wang, L., White, C. S., Okin, G. S., Macko, S. A., and Collins, S. L. (2009). Post-fire resource redistribution in desert grasslands: a possible negative feedback on land degradation. *Ecosystems,* 12, 434–444.

Ravi, S., D'Odorico, P., Zobeck, T. M., Over, T. M., and Collins, S. L. (2007).Feedbacks between fires and wind erosion in heterogeneous arid lands. *Journal of Geophysical Research*, 112, 1-7.

Roig, F., Roig-Junent, S., and Corbalán, V. (2009).Biogeography of the Monte Desert. *Journal of Arid Environments*, 73,164-172.

Roques, K. G., O'Connor, T. G., and Watkinson, A. R. (2001). Dynamics of shrub encroachment in an African savanna: relative influences of fire, herbivory, rainfall and density. *Journal of Applied Ecology*, 38, 268–280.

Rostagno, C. M., and Beeskow, A. M. (2000). Efectos de la erosión en la calidad de los suelos en una estepa arbustiva-herbácea del NE de Chubut. Actas del XVII Congreso Argentino de la Ciencia del Suelo, Mar del Plata, Argentina, 3.

Rostagno, C. M., Defossé, G. E., and del Valle, H. F. (2006). Post-fire vegetation dynamics in three rangelands of North-eastern Patagonia, Argentina. *Rangeland Ecology and Management*, 59, 163–170.

Schlesinger, W. H., Reynolds, J. F., Cunningham, G. L., Huenneke, L. F., Jarrell, W. M., Virginia, R. A., and Whitford, W. G. (1990). Biological feedbacks in global desertification. *Science*, 247, 1043-1048.

Tizón, F. R., Peláez, D. V., and Elía, O. R. (2010). The influence of controlled fires on a plant community in the south of the Caldenal, and its relationship with a regional state and transition model. *Phyton,* 79, 141–146.

Toledo, D. M., Galantini, J. A., Ferreccio, E., Arzuaga, S., Gimenez, L., and Vázquez, S. (2013). Indicadores e índices de calidad en suelos rojos bajo sistemas naturales y cultivados. *Ciencia del Suelo*, 31, 201-212.

United States Department of Agriculture, Natural Resources Conservation Service. (1996). Soil Survey Laboratory Methods Manual, Report 42. Washington D.C.: National Soil Survey Center.

Ussiri, D. A. N., Lal, R., and Jacinthe, P. A. (2006). Soil properties and carbon sequestration of afforested pasture in reclaimed Minesoils of Ohio. *Soil Science Society of America Journal,* 70, 1797–1806.

Villagra, P. E. (2000). Aspectos ecológicos de los algarrobales argentinos. *Multequina,* 9, 35–51.

Villagra, P. E., Defossé, G. E., del Valle, H. F., Tabeni, S., Rostagno, C. M., Cesca, E. M., and Abraham, E. (2009). Land use and disturbance effects on the dynamics of natural ecosystems of the Monte Desert: implications for their management. *Journal of Arid Environments*, 73, 202–211.

Walkley, A., and Black, I. A. (1934). An examination of the method for determining soil organic matter and a proposed modification of the chromic acid titration method. *Soil Science*, 37, 29-38.

Wambolt, C. L., and Payne, G. F. (1986).An 18-year comparison of control methods for Wyoming big sagebrush in southwestern Montana. *Journal of Range Management,* 39, 314–319.

West, N. E., and Hassan, J. (1985).Recovery of sagebrush-grass vegetation following wildfire. *Journal of Range Management*, 38, 131–134.

Whicker, J. J., Breshears, D. D., Wasiolek, P. T., Kirchner, T. B., Tavani, R. A., Schoep, D. A., and Rodgers, J. C. (2002).Temporal and spatial variation of episodic wind erosion in unburned and burned semiarid shrubland. *Journal of environmental quality*, 31, 599-612.

Whitford, W. G., Anderson, J., and Rice, P. M. (1997). Stemflow contribution to the 'fertile island' effect in creosotebush, *Larrea tridentata. Journal of Arid Environments,* 35, 451–457.

In: Rangeland Ecology, Management ...
Editor: Victor R. Squires

ISBN: 978-1-63482-504-7


*Chapter 6*

# WORKING WITH LAND USERS IN MONGOLIA TO ARREST AND REVERSE DESERTIFICATION AND CAPTURE CONSERVATION BENEFITS

***Banzaragch Tsesed*[1], *Chantsallkham Jamsranjav*[2], *D. Bolormaa*[3], *S. Terendash*[4] *and Ts. Sukhtulga*[5]**

[1]Ministry of Environment and Green Development, Ulaanbaatar, Mongolia
[2]Mongolian Society for Range Management, Ulaanbaatar, Mongolia
[3]Administration of Land Affairs Geodesy and Cartography, Ulaanbaatar, Mongolia
[4]Research Institute of Animal Husbandry, Ulaanbaatar, Mongolia
[5]National University of Mongolia, Ulaanbaatar, Mongolia

## ABSTRACT

Since Mongolia's democratization and move to a free market, the country has been grappling with the best approaches to deal with pasture degradation caused by both climate change and lifestyle changes of nomadic herders. International donors and NGOs have implemented community based natural resource management projects with the missions mitigating the effects of pasture degradation and livelihoods of herding families through building capacity. While studies have been done regarding the effectiveness of these community based conservation projects, minimal research has been done to understand how the traditions, values, and culture of Mongolia herders affect the success of these pastureland conservation programs. Thus, one of Mongolia's key challenges today is to find a way to manage pastureland in a sustainable way and to develop sustainable herding strategies to adapt to changing socioeconomic conditions.

**Keywords:** dzud, drought, land tenure, user rights, herders, migration, seasonal pastures, land law

# 1. Introduction: The Setting

Mongolia, with its vast territory of 1.56 million square km, is an upland country with about 85% of its area over 1000 m a.s.l., mostly between 1000 and 1500 m. Mongolia has a continental climate with extreme fluctuations in temperature, both daily and annually. July is the warmest month, with mean temperatures between 15°C in the mountains and 20 to 30°C in the southern semi-deserts and desert. The lowest temperatures are recorded in January, with monthly averages under -15°C and minimum temperatures as low as -40°C (Mongolian Society for Range Management, 2009). Precipitation is generally low, varying within the different ecological zones. It ranges from less than 50 mm per year in the extreme south (Gobi desert region) to about 500 mm per year in certain areas in the north. The average countrywide precipitation is about 230 mm per year.

Of the land area, 8.3 % is covered by forest, 80 % by pastureland, and 1 % is cultivated farmland. Mongolia is divided into six natural zones: tundra; high mountain; forest steppe; typical steppe; desert steppe; and desert (see Figure 1).

A total of 72% of the land, or 112.8 million hectares, is categorized as rangeland which supports over 170,000 herder families. The rangeland falls into five ecological zones, namely high mountain, forest steppe, steppe, desert steppe, desert with markedly different terrain, climate, flora and fauna (Mongolian Society for Range Management, 2008). Land use practices (mainly grazing livestock) vary across these five ecological zones, from nomadic herding in the desert zone to transhumance systems in the more fertile forest steppe. In 2010, livestock herding accounted for 20.6% of gross domestic production and 14% of export earnings. About 18% of the total population in Mongolia, or 366,200 people, rely on herding for their livelihood, which in 2008 comprised 34.8% of the economically active population (Mongolian Society for Range Management, 2009). Rangelands in Mongolia have a very short growing season, limited by cold temperatures and by variable precipitation. Pasture growth begins in mid-May and mostly ends after mid-August. The pasture productivity, dominant species and type, length of growing season and phenology are different for each of these natural zones. The pasture productivity, for instance, ranges between 150 and 1,500 kg/ha increasing from the deserts in the south to the forests and forest steppes in the north.

**Table 1. Pasture area size (million hectares) and livestock distribution in five different ecologicalzones, Mongolia**

| Ecological zones | Pasture area million ha (%) | Distribution of livestock (%) | No. of livestock (Sheep units per 100 ha) |
|---|---|---|---|
| Forest steppe | 24.9 (22.2) | 42.7 | 111 |
| High Mountain | 14.0 (12.5) | 14.0 | 62 |
| Steppe | 30.2 (26.9) | 24.6 | 77 |
| Desert Steppe | 26.8 (23.9) | 13.5 | 32 |
| Desert | 16.4 (14.6) | 5.2 | 20 |

Source: Mongolian Society for Range Management, 2009.

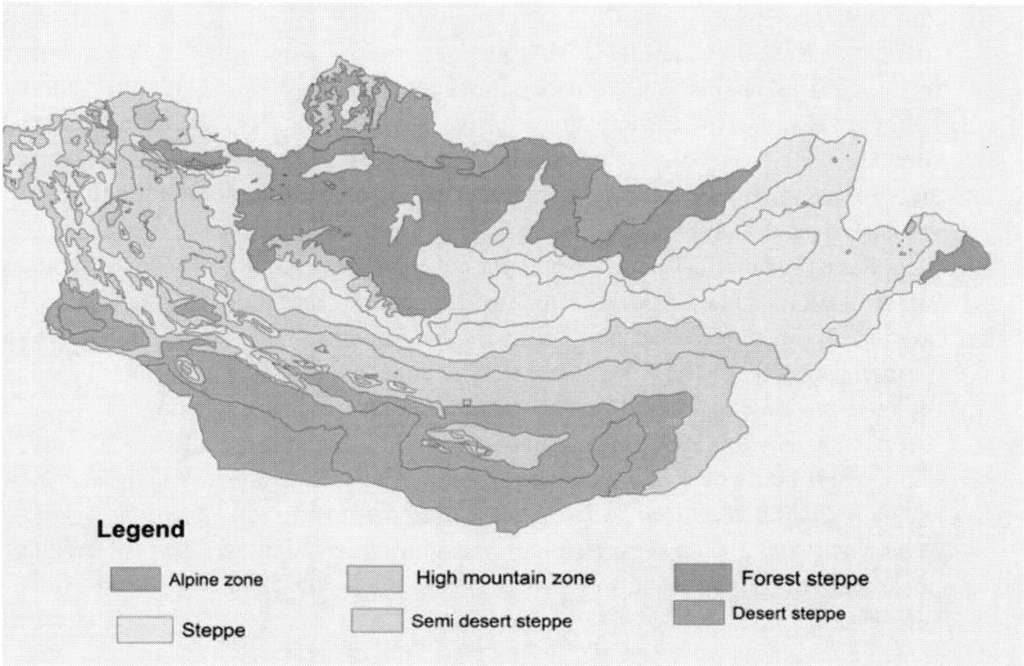


Figure 1. Agro-ecological zones in Mongolia.

The major stages of plant growth are highly correlated to the ecological condition of the growing environment. Plants turn green usually in mid-May in forest steppe zone and dry steppe zone; in high mountains with wet meadow green-up is delayed until mid-June. In the second half of August, most grass species reach their final growth stage and show signs of drying. Various types of pasture have their own peak biomass as the consequence of vegetation composition and conditions, i.e., they differ in productivity. The maximum productivity in forest steppe pastures is 1.1-1.9 t/ha biomass, high mountain area 1.0-1.5 t/ha, meadow 2.9-4.0 5 t /ha, steppe 0.6-1.3 t/ha, and desert steppe 0.3-0.4 t/ha,.

1. Alpine zone covers about 4.5 percent of country's land. The elevation of the alpine zone ranges between 2000-4000 m, covered with few species dominated lichen which are suitable for transhumance grazing during the summer for yaks Figure2/.
2. High mountain zone covers about 3.8 percent of country's land. The plants of rangelands are diverse and major type is grass-forbs type, sedge-grasses type, forbs-sedge type, shrub-grasses type in mountain steppe and meadow.
3. For most of the year, it is cold there and due to insufficient warmth for plant growth, the vegetative period does not exceed for 70 days (Lkhagvajav, 2000). About 90 percent of high mountain zone rangelands is suitable for yak grazing. The herders have divided the high mountain zone into 4 subzones based on elevation and economic significance and have herded the livestock types accordingly. The lowest (*Tsaram*) belt is suitable for yak grazing. The second (*tag*) and third (*saridag*) subzones are not suitable for season-long grazing of the patchy moss and lichen covered vegetation. The highest (*yan*) subzone is the upper limit for plant growth and not used for grazing.

4. Forest steppe area covers 23.3 percent of country's land. The elevation ranges between 1200-2000 m, dominated by pine forest, larch forest, and birch-poplar forest. Grazing land is used primarily for summer grazing by horses, cattle, and large wild herbivores. Forest with an extensive shrub understory is grazed during the summer by all livestock except camels. In riparian meadows located along the rivers and around some lakes. Forest steppe the land is suitable for farming and intensive animal husbandry and horses, cattle, sheep and goats all year round.
5. Steppe area located in central Mongolia covers 25.9 percent of country's land area. The elevation ranges between 550-1500 m. In the steppe area xerophytic species dominate with shrubs such as *Caragana microphylla* and *Artemisia frigida*. The northern portions of the Mongolian steppe along the river basin are highly suitable for cultivation. Rangelands in the steppe zone are used for grazing by horse and sheep, goats and this zone is also famous for beef cattle farming all year around.
6. Semi desert area covers 24.5 percent of land and is dominated by shrubs and forbs such as *Allium polyrrhizum, Stipa gobica, Anabasis brevifolia, Salsola passerina* and *Reaumuria soorgorica*. Rangelands in the semi desert area are used for grazing by horse and sheep, goats and this zone is also famous for camels that graze there all year around.
7. Desert area covers 15.4 percent in land. Sand dunes, gravelly soils and salt marshes are common in this area. Average annual precipitation is less than 50 mm, vegetation is sparse in this zone. Shrubs, sprout during the rainy seasons. Camels and goats are mainly grazed this zone all year around.

## 2. History of Changes in Pasture Use

Knowing the historical patterns of pasture use of nomadic herding in Mongolia helps to understand the present situation and how the herding system might respond to future external socio-economic changes. In the past, herders in Mongolia have been subject to different sets of management rules, depending on the governance structure in place in each period.

During the Mongol State Empire (prior to the 16th century) the territory of the country was divided into four administrative units: Western, Northern, Central and Eastern. Chingghis Khan (1206–1227) was the first to carry out major developments in the relationship of Mongolian nomads to their lands through granting land to his army associates. The army associates were given the right to control the pastures within the boundary of their land grants and the authority to exact labor and taxes from the herders who dwelled there. This marked the first time that groups of herders were associated with specific or fixed territories. After the successful reintroduction of Tibetan Buddhism into Mongolia in 1586, religious monasteries became a dominant political and economic force in Mongolia (Fernandez-Gimenez, 1999). During this time a religious social hierarchy was established and the overall pattern of pasture use was managed by powerful lamas that managed their own territories (Sneath, 2003).

Under the rule of the Manchu, beginning in 1691, the administrative structure of the country was changed. The Manchu divided the 4 *aimags* of a *khalkha* into some 100 military-territorial units or *khoshuu*. The land within a *khoshuu* was under the authority of the prince and controlled by the hereditary nobility, which formed the basis of the so-called feudal pastoral economy of pre revolutionary Mongolia (History of Mongolian People's of Republic, 1981). But under this colonial regime the princes and their subjects alike were forbidden from leaving the area of their birth *khoshuu* (Fernandez-Gimenez, 1999; Riasanovsky, 1965). However, among herders, the *khot ail* was the primary unit of social organization that cooperated in livestock and pasture management issues (*Khot ail* were endogenously developed herding camps, usually kinship-based and the formation of *khot ail* depended on the property relations of its members. Membership in a *khot ail* was flexible and varied from year to year, season to season and usually consisted of 3–5 family units in the most fertile *Khangai* region in contrast to 1–2 families in the desert region. *Khot ail* households cooperated on labor-intensive tasks such as haying, making felt and seasonal movements. In general, it was a kind of social safety net, allowing poor households to benefit from the assistance of wealthier households in informal exchange for their labor (Davaadorj, 1989; Fernandez- Gimenez, 1999; Upton, 2005, 2008).

Between the decolonization period under the Manchu in 1911 and prior to the revolution of 1921, monasteries had an important role in directing the movements of herders and in allocating and controlling pasture use in specific parts of the *khoshuu.* The patterns of land use were managed by a combination of formal regulation imposed by the ruling noble or Buddhist lama (Sneath, 2003), and informal norms, customs described by herders as 'unwritten law'. An overall mixture of formal and traditional regulation governed the use of pasture (Fernandez-Gimenez, 1999).

The Mongolian Republic was founded in 1921. In the early communist period about 300 *soums* were established as territorial/administrative units and the *khoshuu* was abolished. From 1924 to 1990, Mongolia operated under a Soviet–influenced socialist government with a centrally planned economy (Mearns, 1993, 2004). Between 1929 and 1932, livestock were confiscated from nobles and the monasteries and given to the poorest herder households. As a result a pastoral economy was adopted by household subsistence-based small, low productivity, mixed family herds[1]. The government attempt to collectivize all animals in the early 1930s was unsuccessful. Herders opposed taking back their livestock and this resulted in a sharp decline in livestock numbers. Herders slaughtered their animals or drove them into China to avoid giving up their animals. During this pre-collective period from 1921 to 1959, the religious institutions, which were formerly important in pasture use, had been abolished and the influence of the new government was weak. In the absence of the pasture management of the religious feudal system, each group of herders resolutely believed that its own methods and pastures were best suited for its own needs (Fernandez-Gimenez, 1999; Weldon, 2013). During this time, the average number of livestock was 23.2 million head per year, ranging from 20 to 27.5 million (Mongolian Society for Range Management, 2008).

The seasonal movement for pasture use was regulated within the region by periodic moves (see below) and species-specialization in livestock production at the herder camp level (Humphrey, 1978). A large number of mechanical (wind powered) and motorized wells were developed. Moreover, in some areas water for domestic and livestock use was trucked to replenish tanks in remote pastures. Around 65% of the total area of Mongolia had water for livestock and domestic use and this in turn improved distribution of livestock across the

[1] In forest steppe, steppe, high mountain regions, the permitted number of privately owned animals per household was 75 head, whereas in semi-desert and desert regions it was 100 head per family (History of Mongolian People's Republic, 1986).

landscape. A campaign encouraging herders to construct roofed wooden shelters to protect livestock from the danger of severe winter and spring months resulted in more herders returning to the same winter and spring locations every year. Collectives provided assistance by trucks for seasonal movements of households (Fernandez-Gimenez, 1999). In general, as a result of state collective social services the herders' livelihood, education and health status improved. Thus, herders who met the production target for state collectives got paid more and there was no wealth differentiation among the herders (Ressel, 2005). Huge investment, absolute state control and regulation of pasture management in the herding sector changed a nation of nomads into an agricultural industrial state and as a consequence traditional/informal institutions of pasture use were weakened (Mongolian Society for Range Management, 2009).

After the transition from a centrally planned to a market economy in 1990, the state collective livestock was privatized completely by 1993, but pasture land remained state owned. Herders became entirely responsible for their own herd management decisions, as well as production, risks and inputs (Fernandez-Gimenez, 2002). Thus the yield-focused production of the state collective system returned to family-based, subsistence-oriented herding with mixed types of animals (Mearns, 2004). The privatization of state collective livestock provided an incentive for many Mongolians to turn to herding and the livestock sector experienced a sharp upturn. Many newcomers were workers of collectives in the *soum* center, e.g., those who lost a job but were provided with some livestock as a share of the former collectives. In addition, many of those who became unemployed due to the closing of factories in the city migrated to rural areas and became herders (Schmidt, 2006).

As rural communities swelled in size, they also became more heterogeneous. The number of herder families doubled such that in 2007 there was a total of 172,000 herder families and the number of livestock increased from 23 million to more than 40 million between 1993 and 2007 (National Statistical Office, 2007). Many newer herders joined the sector as a social safety net in the wake of de-collectivization and they had less knowledge and skills than those whose families had been herders for generations (The gap between rich and poor herders started to increase in the years between 1993 and 1995) and since 1995 this gap has increased even more. One of the wealth indicators of herders is the number of livestock, which is related to their income. The statistical data from 2007 shows that the number of herder households owning only a few animals has increased; 46.7% of the herder households have less than 50 head of livestock and own only 11.5% of the total number of animals. A total of 35.1% of herders own more than 200 animals or 71.6% of the total livestock in the country (World Bank, 2006). Weakened traditional regulatory institutions, the detrimental effects of newer herders who have less skill and knowledge about herding practices, coupled with free and uncontrolled access to resources due to weak and unclear formal regulations have in practice given herders the freedom to move anywhere. This has increased "trespassing" and out of season grazing of reserved winter and spring pastures, that has resulted in more competition and more frequent conflicts, and has converted the herding system from a controlled pasture system to an open access system leading to overgrazing, a classic example of the tragedy of the commons (Hardin, 1968). The open access system introduced with the adoption of a free-market economy combined with the increasing livestock numbers due to lack of markets or marketing opportunities has resulted in significant deterioration of pasture land and the system is no longer sustainable.

A key herding practice in Mongolia is seasonal movement (see below) where herders set aside pasture as a reserve for the non-growing harsh winter time. These areas are left ungrazed during the summer and autumn growing season. Before 1990 prevention of out-of-season grazing and protection of reserve pasture was enforced both by traditional practices and formal regulation. Increased livestock numbers combined with weakened traditional regulatory institutions during the collective period, and without any clear formal institutions to regulate the pasture use, led to a situation where herders' expectations were not consonant with each other's behavior and this has resulted in an increase in out-of-season grazing and trespassing.

The open access system introduced with the adoption of a free-market economy system, combined with increasing livestock numbers, has resulted in significant deterioration of pastureland and the system is no longer sustainable. Strengthening herder and local government capacity for collective action therefore appears to be the only possible approach to control and reverse the pasture degradation and sustain or improve basis of rural livelihood. The question of the best approach to do this is therefore at the core of the sustainable livelihood issue. Thus, one of Mongolia's key challenges today is to find a way to manage pastureland in a sustainable way and to develop sustainable herding strategies to adapt to changing socioeconomic conditions.

## 3. Current State Role in Land Management

Administratively, Mongolia is now divided into twenty-one *aimags* or provinces, each of which is sub-divided into *soum* (medium administrative unit) or sub-provinces, *soums* sub-divided into *bag* (smallest administrative unit). Each administrative level is headed by a governor who is nominated by an elected assembly-*khural* and formally appointed by the governor at the next-highest level. *Aimag* governors, for example, are appointed by the prime minister, while *soum* governors are appointed by the *aimag* governors. At the *aimag* level, *aimag* governors and *aimag* governments are responsible for defining local policies and implementing government policies. The *aimag* services transfer government tasks to the *soums*, which themselves are barely able to serve at *bag* (smallest administrative unit) and herders' level, while *soum* governors also have the authority to define policies and implement specific activities for the *soum* (Sneath, 2003). Since herding collectives, which were formal regulatory institutions for allocating and managing pasture use, were dismantled, the weakened traditional institutions have been unable to fill the void (Fernandez-Gimenez & Batbuyan, 2004).

In this situation the *Land Law*, which was passed in 1994, authorized user-right contracts (leases) over pastoral resources such as campsites and pastures. The actual implementation, leasing of campsites, began in 1998 (Sneath, 2003). Campsites have been allocated, but pasture was not. However these point locations entail an implicit right to pasture around the winter camp site (Sneath, 2003). According to the Land Law, *soum* governments have the power to control and regulate *soum* land, to allocate possession and user rights to citizens and to impose land fees on land possessors and users, while *bag* governors and *bag* public assemblies are in charge of regulating common land use, making seasonal schedule, allocating hay making areas and allocating winter and spring campsites not already possessed by others (Fernandez-Gimenez & Batbuyan, 2004). But because of incomplete political, and almost no financial decentralization (Mearns, 2004), combined with lack of manpower, skills and sharing of some responsibilities with civil society institutions, the implementation of the Land Law is very weak at the local government level (Mongolian Society for Range Management, 2009). Today, each *soum* government has only one land officer and livestock specialist. The specialists cannot serve herders and work mostly at the *soum* center on reports and papers. Some of them have at the same time the duties of an environmental inspector. The way that the specialists work leads herders to think that the officers perform no useful role. Already 19 years have passed since the collapse of the state collective system, yet still, in practice; the ability of the local governments to fulfil their obligations is limited by lack of manpower and budgetary constraints. There is no extension organization to increase herders' knowledge and or make new technological improvements available to them. Only private units of veterinary services at the *soum* level provide some services through state subsidies.

## 3.1. Technological Specifics of Pasture Use

The specifics of Mongolia's physio-geography, climate, rangeland plants growth and development and monthly and seasonal changes of yields are reflected in the herders traditional pasture use technologies. Herders have got a traditional knowledge of managing rangelands during four seasons of the year while migrating whole year round or in some seasons, which has been the way to preserve rangelands ecological balance and avoid natural risks to animal husbandry (Figure 3)

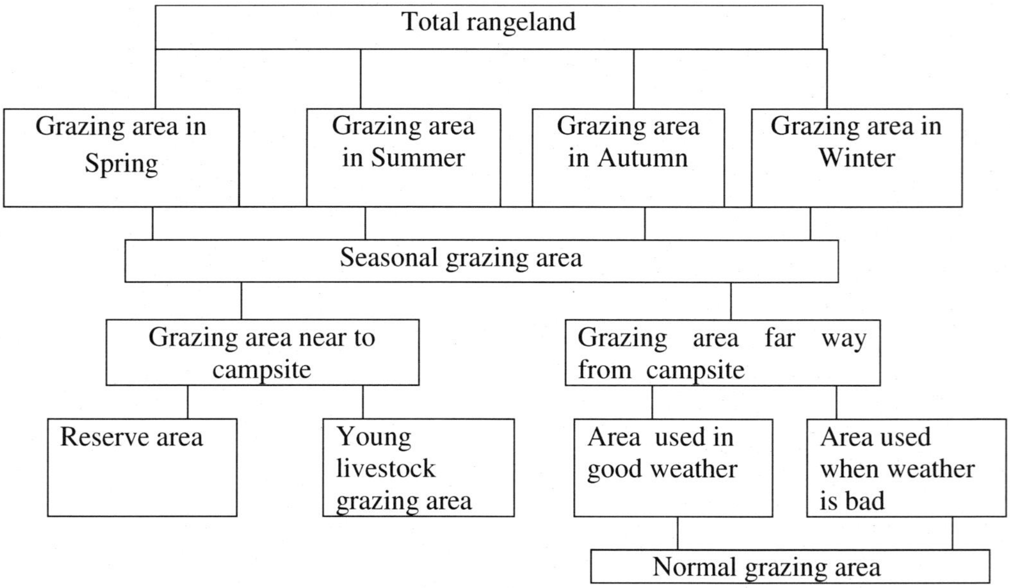


Figure 2. The technological scheme of pasture use by Mongolian herders.

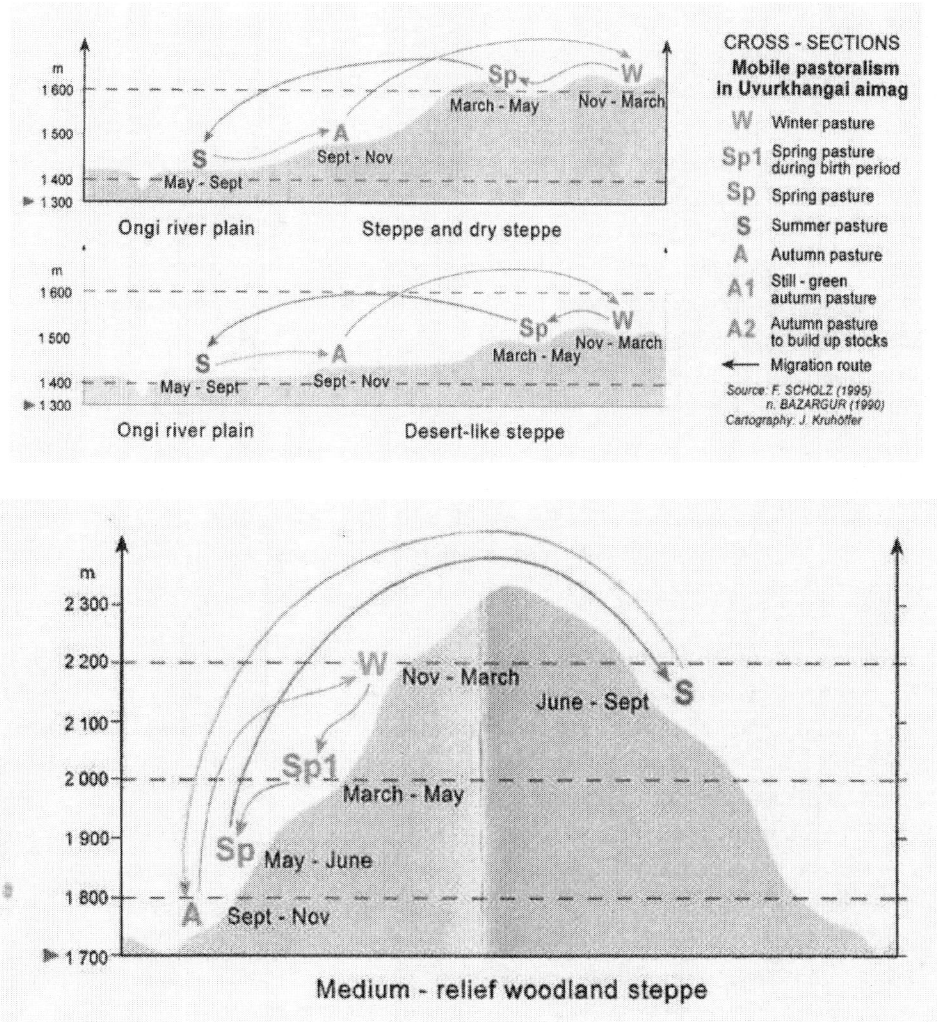


Figure 3. The utilization of season movements in geographic belt.

The Mongolian herders herding style differs from any other nomadic lifestyle around the world with its flexibility to move not depending on the season and to follow the principles of mutual respect while using the rangeland. Every herder has its grazing land for four seasons. During warm summer season the high mountain rangeland is used for grazing and the lowlands for fall and winter seasons grazing. It is important to graze some types of pastures during only some periods of summer.

The herds are migrated from one to another season pasture and four seasonal migrations are referred as four major movements. According the Prof. Bazargur and other researcher's models there are some types of seasonal movements (Figure 5).

Upon arrival at the seasonal grazing lands, the herder divides the pasture into long and near distance pasture depending on relief of land, state of yield, sunshine and water and mineral availability. The more remote pasture is used during suitable weather conditions and near one is during snow, rain and windy circumstances. The remote pasture is 3 km or more and the near one is within a 2 km radius. The near one is also divided into reserve and offspring pasture. The reserve pasture is usually a sheltered area to protect animals during bad weather. The technology to use the pasture varies from season to season. During cold season, a track to grazing areas is created on four sides and herds follow that track to go and return from grazing. This prevents trampling the pasture and helps to preserve reserved grazing areas. The winter pasture is more stable as compared to other seasons and the snow provides water for animals. The time of entry and exit to four season grazing areas differs from place to place. As spring and summer pastures can deteriorate quickly, continuous grazing of is not recommended during warm seasons. Therefore regular rotation and migration to new areas takes place. Depending on the natural and economic specifics of the areas, the seasonal rangeland distribution varies in different parts of the country. The seasonal pasture can be located from tens of kilometres (km) to hundreds of km due to geographical condition, relief and plant cover. The number and distance of grazing areas vary from zone to zone and in forest steppe zones herders move 2-3 times a year and about 30 km, while in *gobi* zones 8-12 times and up to 40-80 km and the steppe herders move 6-10 times and up 10-30 km (see Figures 3, 4, 5). The steppe where the yield is high, water, mineral and pasture is rich (Kherlen river), the pasture is divided for cool and warm season use and the pastures are high yielding and more stable while in the Western parts of the country at least 4 movements are required annually.

One of the common methods in Mongolia to balance the pasture capacity is a long distance grazing *(otor)*. The long distance grazing is done in the less grazed areas or in remote areas used only in special seasons or during drought or severe winter seasons in search of better grazing areas. As a result of such operations, the balance between herd number and pasture's carrying capacity was kept stable.

Today, the number of movements has decreased and in some high mountain areas the herders tend to be grazing near their winter camp whole year around. The pasture that used to be grazed only in one specific season is now used for many days and seasons of the year with high pressure. Thus, the tradition of rotational grazing is lost (Figure 5, Batbuyan et al., 2013). Although many reasons such as lack of manpower and transportation, wish to be near market and other social services or wish to protect their winter camp is named, the ecology is affected negatively.

So, the rotational use of rangeland through migration is the basic tool the herders have followed. There are 3 types of migration in regard to pasture use.

a) Longitude line type. This is a migration through natural zones in search of better pasture. It included the areas of current Uyanga, Saihan-Dulaan and Baruunbayan-Ulaansoums of Uvurkhangai aimag.
b) Elevation line type. This occurs in high mountain areas depending on altitude. The elevation line type migration was popular in Khuvsgul-Zavkhan and Orkhon-Kharkhorin zones but during the early 21st century years it has lost its primary essence.
c) Near water point migration. It is popular in steppe and steppe-desert zones. The water point is the determinant of livestock distribution and time for grazing.

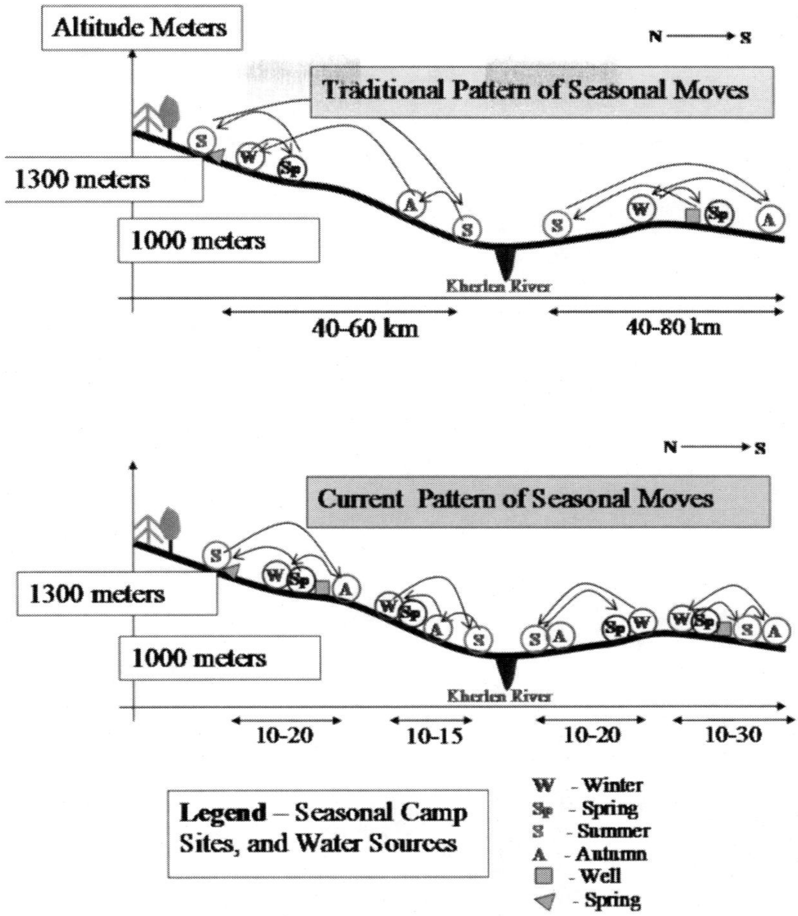


Figure 4.Current changes of seasonal movements.

## 3.2. Current Condition of Pastureland and Changes in Pasture Use

The dry climate makes Mongolia susceptible to drought and rapid depletion of natural pastures (Mongolia Environment Monitor, 2003). It is generally agreed that Mongolia's grassland is sensitive and vulnerable to climate changes and inappropriate management of these ecosystems (Mongolian Society for Range Management, 2009). An understanding of current pasture condition is important when analyzing the present pasture resource use state and helps in the development of an appropriate pasture management plan. A dramatic *decrease* in the productivity of the rangeland was reported by herders and researchers (Dorj et al., 2013). The pasture productivity has decreased by 28.6% in the desert area and by 52.2% in the steppe area from 1961 to 2006 (Figure 6). And at the same time plant diversity was also reduced (Figure 5)

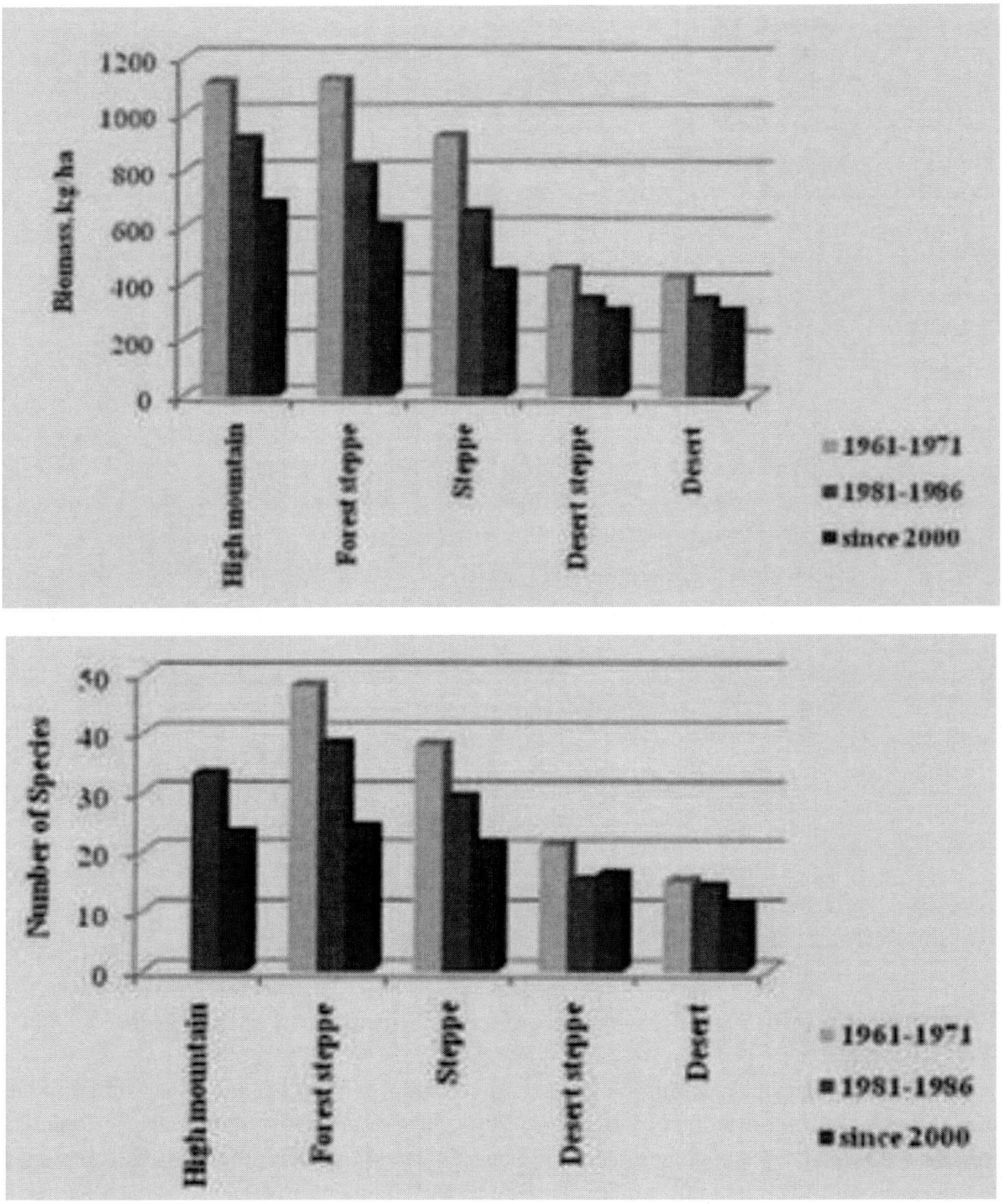


Figure 5. Biodiversity and biomass production is declining.

## 3.3. Climatic Fluctuations and Pasture Carrying Capacity

There is an ecological potential to sustain 86 million sheep unit in Mongolia if water supply is made more reliable. However, as rangelands yield fluctuate from year to year depending on climate conditions, it is important to consider the differences between mild, normal and dry years while estimating the carrying capacity which is crucial to livestock husbandry development. For instance, during mild years about 10 percent increase and during drought and *dzud* 30% decrease should be considered and herd numbers need to be managed accordingly. During recent years, the herd number has increased reaching 47 million heads of animals. However, due to shortage of rangeland and feed, more than 10 million heads of livestock perished in 2010. Despite the severe climate conditions, the high concentration of livestock at some places causing shortage of pasture and feed was one of the main reasons of livestock loss (Figure 6).

Figure 6. Heavy stock concentrations and lack of forage lead to widespread deaths of the livestock and great economic losses to the herders.

The other factor is a climate change. As result of global warming, the climate of Mongolia has also warming impacts and the probability of drought and *dzud* is increasing due to low precipitation and increasing dryness. The global climate change has accelerated the desertification and the boundary of desert areas is expanded by 3.4 percent (Adyasuren 2002). During last 54 years, ending 2000, 36 cases of unfavorable years have been recorded where 6.9 percent of total livestock inventory perished and during 11 normal years about 3.2 percent animal loss recorded. The major factor for animal loss was the feed shortage during the cold, dry winter that was followed by summer drought. About 41.3 percent of Mongolia's rangeland is in a vulnerable natural zone. Biomass yield volume is fluctuating from year to year depending on natural factors like drought, precipitation thus causing difficulties to foresee. An unstable yield is the main cause for limiting the livestock number and therefore it is important to foresee the drought and other natural factors to estimate the pasture's carrying capacity. This has to be the ecological core of the national policy regarding the livestock husbandry.

# 4. Policy Regarding Rangeland Management

One of the conditions to prevent, preserve and restore the rangelands is the development of proper management. The main strategy for pasture management may still be the flexibility offered by migration during four seasons of the year and grazing different type of animals in different types of rangelands. However, there is a conflict between existing old style use of rangeland and the current social development. As rangeland remains a property of the State and livestock are the property of the herders, there is need to let herders understand the value of rangeland and preserve it while investing into it themselves.

The herders can do two types of investments to improve rangeland. The first investment is to keep herd size in accordance with its carrying capacity. In other words, it is a management tool to keep the animal size that can leave behind no less than 50 percent of yield of pasture and let it recover by themself. This does not require much financial means. The second investment includes operations such as fencing the area, planting, fertilizing, and digging wells that require financial means. Any types of investment benefits depend on the level of right to pasture use. It can be more beneficial if such investments made available when herders do it within the group or communities. Although rangeland has been a state property it was used as common land. The theory of "Tragedy of Commons" claims that the management can be only improved when the tenure is secure. The Mongolian traditional system of following natural boundaries over the use of rangeland does not mean that the legal environment over the user rights to grazing is solved (Hannam 2012,). If the herder's rights over the rangeland possession are secured, the herders may carry out many operations, make decisions together making investments to protect, migrate, and build camps etc. joining together based on geographical proximity. While introducing such community management, that the both parties the local government and the herders equally can participate in sustainable pasture management. Therefore there is a demand to renew the legal articles over the rangeland in regards to new social-economic policy and strategy and to adopt new legal environment to coordinate rangeland possession and use framework.

## 4.1. User Rights to Rangeland

Granting of 'User rights' to specified rangelands is one option to regulate access to winter and spring rangelands. However, there must be an ecological and quality assessment made prior to entry to the various designated seasonal pastures. It would be important to certify the natural boundaries that already exist in some areas. When the land is mapped it is important to have a contract made with articles that include baseline data about plant composition, yield, palatability and the soils data that can be the basis for further monitoring and assessment.

One of the tools to coordinate the pasture use can be a pasture use tax. The pasture use fee can be different for each species and for each pasture units. The different types of animals have unequal negative impacts to pasture. The goat and horses are the most destructive species. Therefore the fee must be set higher for these species. The pasture use fee may gradually have positive impacts for improving herd composition and protecting the environment. The pasture's basic value can also be different in each zones and areas. The pasture fee is essential both economically and ecologically. An important point to consider here is to lower the fee if the distance from the settlements goes further as massive migration to central market and to areas with milder weather conditions is continued. The establishment of "Pasture risk fund" based on some portions of pasture fees can contribute and spent for protecting and improving rangeland health, establishing water points, combating rodents and providing training to young herders.

Nowadays, the pasture improvement and the protection operations strongly depend on knowledge and skills of herders. Herders are younger generation and they tend to obtain information from many different sources to manage their herd. However, as they lack complex information regarding environment and the ecology, some operations are not adequately carried out to manage, protect and restore their grazing areas. Therefore, it is important to provide (both indoors and field-based) various information and training particularly about pasture use, assessment, determination of carrying capacity, restoration and hay and feed making technologies, application of annual pasture management plans and implementation, rotational and resting practices both indoors classes and distance education. Only herders with good knowledge of ecology can prevent their land from deterioration and properly manage and use the area.

### 4.2. Establishment of a Rangeland Management Structure

It is crucial to have a pasture management structure both vertically and horizontally coordinating issues regarding proper use, restoration, preservation balance the carrying capacity and prevent from deterioration. However, today the issue of pasture management is divided between multi ministries and agencies and it is not clear who coordinates and monitors the pasture (Hannam, 2012).

At this new socio-economic transition period, the State has to have a policy on rangeland that coordinates the legal rights over the rangeland, prevent and protect the pasture from natural and human negative factors, establish information database and quality network as well as management over the rangeland. The pasture management structure can be operational for multi-purposes that include activities like carrying out extension services to introduce better management technologies regarding assessment of resources and use, to survey, assess and map of deteriorated rangelands for each *aimags*, to plant abandoned croplands, to develop technologies to improve deteriorated pastures and to establish a gene bank for plant seeds etc. Also, the pasture monitoring for assessment, the elaboration of pasture management plan by *aimag, soum* and groups and families are important part of the management system.

## 5. Common Pool Resource Systems and Herder Institutions

Regulating access to Mongolian pastureland and adapting existing and instituting new rules among herders is a very important but complicated issue in the current changing socio-economic situation. Thus, rethinking of the common pool resources system itself and reflecting on experiences of commons in other countries dealing with resource use problems will be helpful for Mongolia.

Clarifying the concepts related to resource systems and those concerning property rights helps to understand common-pool resource problems better. Common-pool resources include natural and human constructed resources (Ostrom et al., 1999). Examples of common-pool resources include irrigation systems, fishing grounds, pastures, forests, water and the atmosphere. A pasture, for instance, allows for a certain amount of grazing each year without the core resource being harmed. In case of excessive grazing, however, the pasture may eventually degrade, leading to less pasture productivity, which then leads to soil erosion. If common pool resources are managed carefully the use can be extended, whereas without limiting and regulating the optimum amount of use and harvest -ing, this resource can easily be damaged, overused or polluted (Sterner, 2003). Property rights arrangements have been classified into four broad types in relation to common-pool resources (Ostrom et al., 1999).

- **State property**; ownership is by national, regional, or public agency that can forbid or allow use by individuals.
- **Private property**; individual has the right to exclude others and regulate resource use.
- **Common property** has an identifiable group of users who share resources according to their own mutual regulations and who can exclude others.
- **CPRs** (common property regimes) with open access, and uncontrolled access to use of resources.

In practice no single property regime works completely efficiently, sustainably and fairly in relation to all users. Since Mongolia shifted toward a market economy, the State has lacked the capacity to manage the private livestock of herders on its land. This mismanagement has led to unsustainable practices of pasture land use or open access and resource degradation (Mearns, 2004; Sneath, 2003). This in turn has created problems for the government, which will then have difficulties in supporting the herding families whose livelihood completely depend on this resource. In an efficient market economy, private ownership of economic assets is promoted, enabling government to collect taxes, allowing land owners to obtain credit using land as collateral, and stimulating land improvement that leads to increased productivity and land stewardship (Sterner, 2003). This logic will be difficult to apply to the pasture land supporting the mobile herding system of Mongolia. In this climate, Mongolia might be able to learn from the experiences of other countries, especially about managing common grazing land successfully.

The approaches of many countries (e.g., China, Iran) to nationalize ecosystem properties, taking land as State property in order to prevent degradation instead of managing it as a common pool resource, has often failed. Governments have often lacked both human and financial resources to effectively control and manage state-owned land. Another reason for their failure in sustainable land management is that by nationalization they fail to recognize and respect local traditional rights.

In many cases privatization of the right to use natural resources does not provide an appropriate solution. For example, when there are potential environmental hazards from e.g., pollution from several separate but neigh-boring entities, assigning responsibility, liability, and rights becomes problematic. Furthermore, the social welfare aspect is problematic in private property regime. Besides ecological concerns, social issues may arise if all the financially better-off people manage to buy all the resource shares. Poor people have little property and mostly it is they who benefit from common or open access systems. Supporting private property will widen the wealth gap, making the poor opposed violently to private property (Sterner, 2003). When resources are getting scarce due to overuse, then people may find ways to collaborate actively and develop a common property regime (CPR) to maximize their benefits. Through complex community norms and consensus decision making the common pool resource can be protected and used appropriately. Individual behavior however, varies depending on the situation and circumstances. Generally there are three common behaviors of users of a CPR: (i) Some behave in a narrow self-interested way, never cooperating but rather acting as free riders,( ii) while others are selfish in certain situations only and iii) thirdly, others who are willing to initiate reciprocal cooperation in the hopes that others will return their trust (Ostrom et al., 1999). So devising the rules to control the access to the resources and institute rules to resolve conflict between individual and collective rationality is a difficult task. Setting the limits too high would lead to overuse and eventually to the destruction of the core resource, while setting the limits too low would unnecessarily reduce the benefits obtained by the users (Sterner, 2003).

The "Tragedy of the Commons" occurs when there are no constraints on accessing the resources. There is now wide acceptance that managed commons, as compared to open access/unmanaged commons, do not have to be degraded and may be an effective method for managing resources sustainably (Williamson, Brunchkhorst & Kelly, 2003).

Communal management of grazing land is important for long term ecological and social sustainability. Since all herders in Mongolia have a similar herding lifestyle and want to have better pastureland and a better livelihood, mutually agreed institutional building based on traditional arrangements for pasture use is most needed and a crucial step to bringing needed management to the presently unmanaged situation. Private pastureland ownership demands precise management from the owner even in areas where such lands are properly administered e.g., Australia

## 6. Initiatives towards Sustainable Pasture Use – Herder Group Approach

A three-year consecutive natural disaster (a *dzud*, in this case severe snow storms, cold and drought) from 1999 to 2002 resulted in national livestock losses close to 30% and significant loss of livelihood (Figure 9).

Many of the less experienced herders, those who joined the herding sector later, lost most of their animals and became part of the marginal group of poor (Mau & Chantsallkham, 2006). During these times, herders came to understand that cooperation is crucial to overcome harsh times and to improve their livelihood. Herder groups therefore evolved spontaneously in Mongolia throughout the late 1990s, in response to the need of herder communities to survive under changing ecological and socioeconomic conditions or driven by wish to improve their livelihood (Schmidt, 2006).

Figure 7. Harsh winters and lack of feed are major causes of livestock deaths and under severe dzuds up to 30% of the livestock inventory may be lost.

At that time the Government requested assistance from donor organizations to lessen the harm of *dzud*, to reduce the poverty, improve the livelihood of herders and reduce the environmental degradation. Most donor organizations provided their support in the form of training, skill development and provision of low-interest micro-credits to herders through herder groups (Mau & Chantsallkham, 2006). Donors saw the benefits of working with herder groups to reduce transaction costs and increase efficiency to provide support and involve herders who live far from each other. As a result, about 15,954 rural families organized into 1,957 donor supported herder groups. Where 206 herder groups are considered to be a business cooperative (11%), 355 herder groups are registered as NGOs (18%) and 1396 are classed as informal groups (71%) (Mau & Chantsallkham, 2006).

In general, the main impact of small herding group evolvement in herder communities revealed that they were lacking in formal institutions to coordinate them. Herders also understood that they could reorganize themselves into collective action to improve livelihood to fill the vacuum left by the abolition of the state collective system. The experience of the last decade showed that relatively small voluntary herder groups could take over income generating activities and play other roles in the herding community but they have not had enough capacity or coordination skills as an organization to assume a leading role in addressing the crucial issues of management of the pasture land (Chantsallkham, 2011).

## 7.1. The Pasture User Group (PUG) System

The rangeland specialists distinguish two different ecological systems in rangelands (Box 1).

**Box 1. The Non-Equilibrial Nature of Rangeland Ecosystems**

Two concepts have dominated the thinking regarding ecology and land use: (1) that ecosystems are equilibrium systems and (2) that people are an outside source of disturbance. Modern research in ecosystem function is showing that ecosystems are not necessarily near thermodynamic equilibrium, and that people are integral constituents of ecosystems. Most pasturelands are non-equilibrium systems which can behave in ways that appear strange in relation to what is expected from an equilibrium system. For example, when the source of disturbance is removed from an equilibrium system, it is expected to return to its previous state. When non-equilibrium systems are disturbed, however, they are likely to adopt an entirely new stable or quasi-stable state.

Another characteristic of non-equilibrium systems is that they may oscillate continually due to their own internal dynamics. These oscillations can be regular (limited cycles) or irregular (chaotic oscillations).These characteristics of non-equilibrium systems have important management implications. Significantly, they do not try to rehabilitate themselves as they do not necessarily return to their previous state when the source of disturbance is removed. Moreover, in most cases it might not be possible to return them to their original condition. Furthermore, it also means that one management regime will not suffice for all sites, even at the local level. The previous history and present disturbance regime will determine the outcome of management systems. This also means that levels of human investment (capital, labor and knowledge) are major determinants of the behavior of managed dryland systems. The lesson is that no single set of general management prescriptions can be offered for the 'proper' use of pasturelands. The level of human investment, the objectives of the human society, and the characteristics, as well as the previous history, of an area will determine which management regimes satisfy human aspirations and which do not.

The *equilibrium* system defines areas with more or less regular climatic conditions, where normally the quality of pasture land can be controlled by the number of livestock. In *non-equilibrium* systems highly variable climatic conditions are more important for the quality of the pasture than the number of livestock (Heshmati and Squires, 2009). These are areas in which movements of livestock cannot follow regular patterns and have to be organized according to the availability of fodder, which is different from year to year (an opportunistic approach). Pasture User Groups (PUG) have been successful in the mountains, forest steppe and steppe regions of Mongolia where equilibrium pasture systems of mixed equilibrium- disequilibrium systems prevail, rather than the disequilibrium pasture systems of the desert steppes or desert areas.

PUGs consist of the traditional herders in a particular territory (often within a *bag*) who through customary law have the exclusive right to manage the pasture area in their traditional grazing area. They are inclusive groups and no traditional herder in the area can be excluded from participating. PUGs on their own can facilitate seasonal rotations, resting of pastures and other technical means of pasture management, and incorporate these into Pasture Management Plans (PMP). They can also facilitate and negotiate seasonal and permanent movement in and out of the PUG area. However, in order to negotiate the definition of PUG areas and boundaries among different PUGs, they need to form an Association of PUGs (APUG) at a higher level, most often the *soum* level, and negotiate the territorial allocations with the support of the *soum* government. Another essential task for PUG pasture land management is setting reserve pasture – *otor* for critical times during drought and *dzud* before defining PUG boundaries. These in turn will have to collaborate with PUGs to set aside and enforce the *otor* areas. One of the problems in reintroducing reserved pasture areas will be that in many parts of the country the level of pasture land degradation and the animal pressure are so high that most years will be considered as crisis years where access to reserved pasture land has to be allowed. This situation rules out a return to sustainable land management any time soon.

## Conclusion

Following the economic and political shifts in Mongolia in the early 1990s, the state herding collectives were dismantled and their livestock was privatized. Pastureland remained state owned, to be used in common by the herders of defined administrative areas (*soum* and *bag*). Herders became entirely responsible for their own herd management. Weakening the customary regulatory institutions of the collective period, without any clear formal institutions to regulate pasture use, led to unsustainable grazing practices. Consequently, excessive pastureland degradation in Mongolia has raised debate over how pastureland should be allocated and regulated in a market economy. An overview of historical patterns of pastureland use and a case study of the territory-based herder group approach in a semi-equilibrium ecosystem suggests that a promising solution to address the current trends in unsustainable grazing practices and increasing conflicts over pasture use and access is residence-based groups with exclusive user rights over pasture within their area. In the past, the patterns of pastureland use were managed by a combination of formal regulation imposed by governance structure in place in each period together with informal norms and customs described by the herders themselves. A combination of formal and traditional regulation should govern the use of common pasture. In the current situation, privatization of Mongolian rangeland does not provide appropriate solutions and is opposed by almost all Mongolian herders. Focus group discussion and informal interviews with the members of territory-based groups and local authorities has revealed that the direct involvement of resource users in management is a promising approach to improved pastureland management and herders' cooperative behavior, thereby altering the open access situation. Strong informal users cooperation is more powerful and could help to establish formal regulation on pastureland and or policy formulation in the future.

The open access system introduced with the adoption of a free-market economy system, combined with increasing livestock numbers, has resulted in significant deterioration of pastureland and the system is no longer sustainable. Strengthening herder and local government capacity for collective action therefore appears to be the only possible approach to control and reverse the pasture degradation and sustain or improve basis of rural livelihood. The question of the best approach to do this is therefore at the core of the sustainable livelihood issue. Thus, one of Mongolia's key challenges today is to find a way to manage pastureland in a sustainable way and to develop sustainable herding strategies to adapt to changing socioeconomic conditions.

Sustainable land management cannot be assured under the present system.

## References


Adyasuren, Ts. & Borjigdkhan. (2002). Land degradation assessment in Mongolia. http://rwes.dpri.kyoto-u.ac.jp/~tanaka/APHW/APHW2004/ proceedings /JSD/56-JSD-M129/56-JSD-M129.pdf

Batimaa, P., Bat, B., Tserendash, Sh. & Myagmarjav, B. (2008). Adapting to Drought, Zud and Climate Change in Mongolia's Rangelands. In: *Climate Change and Adaptation*. Neil L. et al. (eds). Earthscan. UK

Bazargur, D., Shiirevadja, C. & Chinbat. B. (1993). Territorial Organisation of Mongolian Pastoral Livestock Husbandry in the Transition to a Market Economy. Policy alternatives for livestock development in Mongolia (PALD): *A Research and Training*. Project Research Report 1,

Chantsallkham, Jamsranjav. (2009). Sustainable rangeland management in Mongolia: The role of herder community institutions. *Land Restoration Training Programme Final project 2009 Keldnaholt, 112 Reykjavík, Iceland 35 p.*

Dagvadorj, D. (2002). Climate Change and food Security, in Proceedings of the Workshop on Climate Change and Agriculture, 23 October 2002, Darkhan. ed. Mijiddorj J and Davgadorj D., 12-20.

Dorj, O., Enkhbold, M., Lkhamyanjin, S., Mijiddorj, Kh., Nosmoo, A., Puntsagnamil, M. & Sainjargal, U. (2013). Mongolia: Country features, the Main Causes of Desertification and Remediation Efforts. 217-230 In: G.A. Heshmati and V.R. Squires (Eds) *Combating Desertification in Asia, Africa and the Middle East*. Springer, Dordrecht.

Fernandez-Gimenez, M. E. (1999). Sustaining the steppes: A geographical history of pastoral land use in Mongolia. *Geographical Review*, *89*, 315–342.

Fernandez-Gimenez, M. E. (2002). Spatial and social boundaries and the paradox of pastoral land tenure: A case study from post socialist Mongolia. *Human Ecology*, *30*, 49–78.

Fernandez-Gimenez, M. E. & Batbuyan, B. (2004). Law and disorder: Local implementation of Mongolia's Land Law. *Development and Change*, *35*(1), 141–165.

Hannam, I. (2012). International Perspectives on Legislative and Administrative reforms as an aid to better land stewardship. 407-429 In: Victor Squires (Ed) *Rangeland Stewardship in Central Asia: Balancing Improved livelihoods, Biodiversity Conservation and Land Protection* Springer, Dordrecht.

Hardin, G. (1968). Tragedy of the Commons. *Science*, *162*, 1243–1248.

Heshmati, G. A. & Squires, V. R. (2009), New thinking in range ecology, in *Range and Animal Sciences and Resources Management*, [Ed. Victor R. Squires], in *Encyclopedia of*

*Life Support Systems (EOLSS)*, Developed under the Auspices of the UNESCO, EOLSS Publishers, Oxford, UK

Humphrey, C. (1978). Pastoral Nomadism in Mongolia: The role of herdsmen in cooperatives in the national economy. *Development and Change*, *9*(1), 133–160.

Institute of history: Mongolian Academy of Science. (1986). *History of Mongolian People's Republic*. Ulaanbaatar, Mongolia: State Publishing Company.

Lkhagvajav, N. (2006). *Decrease of pasture land productivity. Proceedings of Theoretical and Practical Conference on Issues and Challenges of Pasture Management. UB* Ulaanbaatar.

Mongolia: Munkhiin useg Publishing. (In Mongolian).

Mau, G. & Chantsallkham, J. (2006). *Herder Group Evaluation: A study of herder groups, their present status, and future potential*. Mongolia: UNDP.

Mearns. R. (1993). Pastoral Institutions, Land Tenure and Land Policy Reform in Post-Socialist Mongolia. Policy alternatives for livestock development in Mongolia (PALD): A Research and Training. Project Research Report 3,

Mearns, R. (2004). Decentralization, rural livelihoods and pasture-land management in post socialist Mongolia. *Development Research, 16*(1), 133–152.

Mongolian Society for Range Management. (2008). *Warm welcome to the steppes of Mongolia.* Brochure for the participants of the International VIII Rangeland Congress XXI International Grassland Congress Post Tour in Mongolia.

Mongolian Society for Range Management. (2009). *Livelihood Studies of Herders in Mongolia: Draft study report*. Unpublished internal report. Ulaanbaatar. Mongolia.

Mongolia Environment Monitor. (2003). *Land Resources and Their Management.* World Bank. Environment Monitor. Ulaanbaatar. Mongolia: World Bank report.

National Statistical Office. (2007). *Statistical Year Book*. Ulaanbaatar: Interpress.

Ostrom, E. (1990). *Governing the Commons*. Cambridge: Cambridge University press.

Ostrom, E., Burger, J., Field, C., Norgaard, R. & Policansky, D. (1999). Revisiting the commons: Local lessons, global challenges. *Science*, *284*, 334–338.

Riasanovsky, V. A. (1969). *Customary Law of the Nomadic Tribes of Siberia* L'Homme *9* (4), 121-1122

Ressel, C. (2005). Cooperation strategies in Mongolian pastoralism during the socialist collective economy. *Inner Asia*, *7*, 193–214.

Schmidt, S. M. (2006). Pastoral community organization, livelihoods and biodiversity conservation in Mongolia's Southern Gobi Region. 18-29 In: Bedunah, Donald J., McArthur, E. Durant, and Fernandez-Gimenez, Maria, comps. 2006. Rangelands of Central Asia: Proceedings of the Conference on Transformations, Issues, and Future Challenges RMRS-P-39. Fort Collins, CO: U.S. Department of Agriculture, Forest Service, Rocky Mountain Research Station.

Schmidt, Sabine M. (2006). Pastoral Community Organization, and Biodiversity Conservation in Mongolia's Southern Gobi Region. USDA Forest Service Proceedings RMRS-P-39. 2006

Sneath, D. (2003). Land use, the environment and development in post-socialist Mongolia. *Oxford Development Studies. 1*, 441–459.

Sterner, T. (2003). *Policy instruments for environmental and natural resource management: Resources for the Future*. Washington DC: Resources for the Future.

Upton, C. (2005). Institutions in a pastoral society: Process of formation and transformation in post socialist Mongolia. *Comparative Studies of South Asia, Africa and Middle East, 25*(3), 584–599.

Upton, C. (2008). Social capital, collective action and group formation: Development trajectories in post-socialist Mongolia. *Human Ecology, 36*, 175–188.

Williamson, S., Brunchkhorst, D. & Kelly, G. (2003). *Reinventing the Commons.* New South Wales: The Federation Press.

Weldon, K. E. Y. (2013). Think like a Mongolian: Cultivating community based pasture management. Independent Study Project (ISP) Collection, Paper 1559 http://digitalcollections.sit.edu/isp_collection/1559/

World Bank. (2006). *Mongolia Poverty Assessment Report.* Ulaanbaatar, Mongolia.

In: Rangeland Ecology, Management ... ISBN: 978-1-63482-504-7
Editor: Victor R. Squires 

*Chapter 7*

# PASTURELAND MANAGEMENT IN TAJIKISTAN'S RURAL REGIONS FOR ECOLOGICAL SUSTAINABILITY, ECONOMIC PROFIT AND SOCIAL EQUITY

***Victor R. Squires[1], Barno Kurbanova[2] and A. A. Madaminov[3]***
[1]College of Grassland Science, Gansu Agricultural University, Lanzhou, China
[2]Giprozem, Dushanbe, Tajikistan
[3]Institute of Botany, Dushanbe, Tajikistan

## ABSTRACT

For at least the first half of the twenty-first century, most people in Tajikistan will derive their living or depend on the productivity of the soil, water and vegetation resources (i.e., conservation benefits). Yet Tajikistan, more than all other countries in Central Asia and indeed among the former Soviet Republics, has real physical constraints on the productivity of the environment. These constraints are exacerbated by the dysfunctional governance and administrative policies. These limitations have been compounded by layers of human use and mismanagement and these pressures on the natural resource base and environmental systems have been continuously increasing over the past few decades. Rapid population growth, urbanization, and consumptive lifestyles, rising economic output coupled with an increasing incidence of poverty, have all contributed to Tajikistan's struggle to combat desertification, and arrest or reverse environmental damage in all its forms. Historically, nomadic pastoralism remained relatively constant and, presumably, in balance with natural resources i.e., herders and agro pastoralists were moving up to hundreds of kilometres to follow the natural seasonal progression of rangeland forage production across elevation and latitudinal gradients. These types of grazing were seasonal and plants had a chance to mature before they were grazed again. An important aspect of this type of system was that the 'return time' was long enough for plants to recover their vigor, set seed, and build up root reserves. Although nomadic grazing was seen as 'primitive', making it difficult for government agencies to serve and regulate rural communities, it can actually allow greater productivity than the more 'modern' grazing systems that may be rational over a small spatial scale but are sedentary at the spatial scale of the most relevant environmental heterogeneity.

**Keywords:** nomadism, land tenure, landlessness, forage, constraints, ecological, socio economic, legislation, governance, biodiversity

## 1. Introduction

The Republic of Tajikistan is a land locked mountainous country in Central Asia. Mountains accounting for more than 93% of the country's territory, with mountain peaks rising to 3000-7000 m. Nearly half of the country's territory is situated at elevations of over 3000 m above sea level (a.s.l.). Tajikistan is home to the highest mountain peaks in the Pamirs and Alay ranges. Tajikistan borders Kyrgyzstan in the north and north and north-east, lowlands Uzbekistan in the north and north and west, Afghanistan in the south, and China in the east. The total areas of the country is 142, 600 km$^2$. Figure 1 shows the principal geographical features, including the major river systems.

Figure 1. Tajikistan is a small mountainous land locked country in Central Asia.

Tajikistan has a climate with hot dry summers and relatively mild winters. In the lowlands, in July average temperatures are above 30 $^{0}$ C, while in the mountains the average temperatures range from 6 to 16 $^{0}$C. The frost-free period lasts about 250 days. Winters are warm and short, but penetration of air masses from the north can cause temperatures to plummet to -15$^{0}$C and lower. In the southern areas the average January temperature does not drop below -2$^{o}$C which is why livestock over-winter there. In the mountains the January temperatures are in range of -10$^{o}$-12$^{o}$ In the Pamir climate changes from arid to polar depending on the elevation. Yaks are well adapted to over winter there. In the eastern Pamir, winter temperature can be low as -40$^{o}$C

## 2. Landscapes and Pastureland-dependent Societies

Tajikistan's pasturelands are characterized by immense diversity in climatic regimes, physical conditions, cultural and historical roots, and economic conditions (Strong and Squires, 2012). This diversity presents problems for the manager and the politician, since it requires development of flexible and variable management strategies that can be at odds with national policies and with the interests of increasingly urbanized populations.

Pastureland societies are becoming progressively marginalized in political, cultural and economic terms (Squires and Hua, this volume). Although linked to national economies and through them to world markets, many rural people have little control over their destiny. Many development, land-use and trade policies, for example, are formulated without considering their implications for the people inhabiting Central Asia's rangelands and pasturelands. This is especially so in Tajikistan. Yet, the influence of these policies on land-use can be dramatic. The situation is complicated further by the relatively slow speed at which pastureland ecosystems reveal their responses to policy. The imposition of price controls on meat or other animal products or cross border trade embargos a change in tax policy that alters the relative profitability of alternative livestock, will gradually change pasture species composition and then, some ten to fifteen years later, begin to affect animal productivity and through this the welfare of local people as in the situation with cashmere goats in Mongolia where goat inventories exploded. Until 1990 Mongolia was a satellite of the Soviet Union. Grazing herds were collectivised and nomadic herders (who still make up half of Mongolia's population of 2.5m today) were state employees, with guaranteed pensions and with fixed quotas on the number of animals they could tend. Cashmere wool and meat went mainly to Soviet markets, and the state-owned Gobi Cashmere was the monopoly processor. With democracy and the chaotic introduction of a market economy, the herds were privatised. Impoverished herders, desperate to boost their income, cross-bred their goats with coarser-haired angora ones. The price of raw cashmere has plummeted, to less than a third of its level in the early 1990s. Many people were forced into poverty.

# 3. Determinants of and Constraints to Land Use in Rangeland and Pasturelands

Rangelands and pasturelands are ecosystems characterized in the un-degraded state, by the presence of a continuous herbaceous layer, mostly grasses, and sometimes a discontinuous canopy of shrubs. Although natural grasslands are physiognomically similar, there are major ecological differences among them in soil fertility, rainfall, elevation and aspect. The people occupying such lands differ culturally and in their historical roots and experiences (Masselli and Imam, 2012; Rahminov, 2012). Land use reflects both the natural environment and the human society living in an area. Thus the response of a pastureland to the same policy change can vary considerably. When people extract energy and materials to fulfil their needs for water, food, fuel and fiber, they inevitably modify the natural environment and the potential productivity of the land. It is critical, therefore, to understand the interactions within and between both the natural biotic and abiotic components of the ecosystem and the socio-economic component (Figure 2). The most important elements are land capability, rainfall seasonality, episodic events (frequency and severity) land tenure arrangements and cultural diversity.

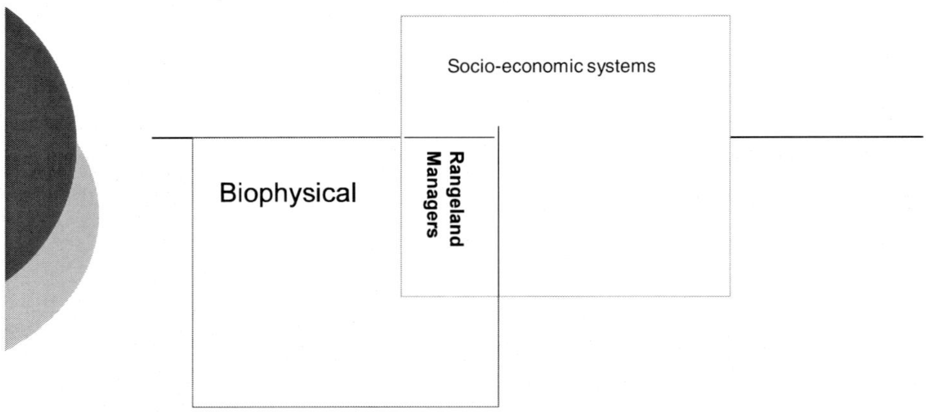


Rangeland Managers operate at the interface between two systems.

Figure 2. Two important systems operate in the rangelands and close attention must be paid to each one.

## Rainfall Seasonality

Climatically the region's pasturelands are characterized by the alternation of a rainy and a dry period. The length of the rainy season, as well as the total precipitation, are major constraints to the successful development of agricultural systems in savannas. The length of the rainfall season varies between three and four months in the driest areas, to between six and seven months in the wetter ones. Absolute yearly rainfall varies between 200 and 1200 mm.

## Episodic Events

All pastureland ecosystems are characterized by the need to manage and cope with episodic events that constrain opportunities for development and for implementation of sustainable land use systems. Periodic drought, extremely cold winters (called *dzud* in Mongolia) floods/landslides and, in some places, wildfire are examples of episodic events that occur within the region.

## Cultural Characteristics

Many pastoral cultures have deep and powerful roots and traditions that cannot be changed easily. Cattle, goats, sheep and other ungulates appear to have been first domesticated some 9000 to 11,000 years ago, and 6000 to 8000 years ago people began to take their herds out onto the pasturelands (Smith, 2008). Sheep dominate in most Central Asia but herds and flocks contain cattle, goats, horses, camels and donkeys, in various combinations.

Figure 3. Much of the lowland area is semiarid to arid and adapted animals like camels graze/browse there. (Photo V.Squires).

With few exceptions, the desire for intensive agricultural development within the region is a recent phenomenon. This, however, does not mean that local people do not produce or eat agricultural products. All but the wealthiest, obtain more than half their food-energy needs from vegetable gardens, orchards and opportunistic cropping and by trading livestock products for wheat and other cereals (Lane and Scoones 1993).

## Beyond Direct Sustenance Values

Well noted for their rich environmental and ecological values that help to preserve ecological diversity (Squires and Safarov, 2013) some pasturelands generate significant tourist revenue. The development of this potential for eco-tourism is still in its infancy in many areas, however, and sometimes hindered by political instability. Often vast landscapes mix monotony with surprise (Figure 4).

Figure 4. Yardangs (sculpted by wind and water over time) are a feature of many rangelands.(Photo V.Squires).

Unlike most cropping systems, pasturelands are non-equilibrial or multi-state systems (Heshmati and Squires, 2010) which tend to flow from one state to another (Box 1).

Traditional nomadic pastoralism fully exploited these characteristics; for example, by moving from one area to another in sympathy with seasonal conditions (Squires, Hua, Zhang and Li, 2010). Significantly, there is mounting evidence that these forms of use are more economically efficient (productive) and less ecological damaging than the sedentary systems that characterize other landscapes (Ellis and Swift,1988). Breman and Traore (1987), for example, have found that annual liveweight gain in nomadic and semi-nomadic cattle is twice as high as that achieved by sedentary grazing. The reason for this is that herders seek to maximize the production of product (milk, meat, blood) per unit area whereas commercial ranchers seek to maximize productivity (wool, beef) per head.

**Box 1. The non-equilibrium nature of dryland ecosystems**

Two concepts have dominated the thinking regarding ecology and land use: (1) that ecosystems are equilibrium systems and (2) that people are an outside source of disturbance. Modern research in ecosystem function is showing that ecosystems are not necessarily near thermodynamic equilibrium, and that people are integral constituents of ecosystems. Most pasturelands are non-equilibrium systems which can behave in ways that appear strange in relation to what is expected from an equilibrium system. For example, when the source of disturbance is removed from an equilibrium system, it is expected to return to its previous state. When non-equilibrium systems are disturbed, however, they are likely to adopt an entirely new stable or quasi-stable state.

Another characteristic of non-equilibrium systems is that they may oscillate continually due to their own internal dynamics. These oscillations can be regular (limited cycles) or irregular (chaotic oscillations).These characteristics of non-equilibrium systems have important management implications. Significantly, they do not try to rehabilitate themselves as they do not necessarily return to their previous state when the source of disturbance is removed. Moreover, in most cases it might not be possible to return them to their original condition. Furthermore, it also means that one management regime will not suffice for all sites, even at the local level. The previous history and present disturbance regime will determine the outcome of management systems. This also means that levels of human investment (capital, labor and knowledge) are major determinants of the behavior of managed dryland systems. The lesson is that no single set of general management prescriptions can be offered for the 'proper' use of pasturelands. The level of human investment, the objectives of the human society, and the characteristics, as well as the previous history, of an area will determine which management regimes satisfy human aspirations and which do not.

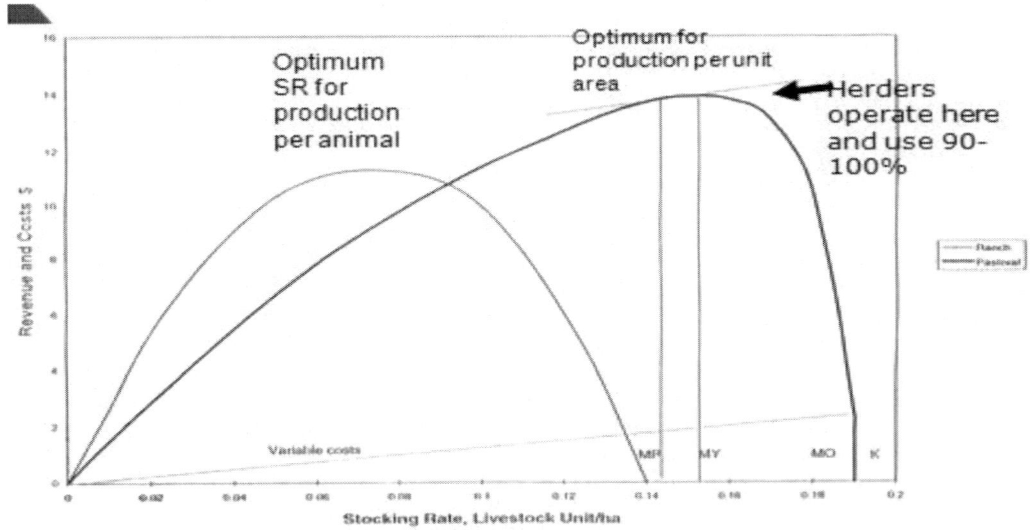


Figure 5. there is a big difference in the thinking of commercial livestock producers whose aim is to maximize productivity per animal and the traditional herders whose focus is on production per unit area of land.

# 4. Current Issues

*Population pressure* Tajikistan's population rate has been falling but still sits at about 2% per year. The population in 2015 is expected to rise above 8.6 million. The Age structure shows a preponderance of people under 64 years of age but those aged 0-24 years represent 34.3%. The proportion who live in the rural areas is over 70% at a population density of about 50 per $km^2$. The area of arable land per capita (about 0.1 ha) is low and shortage is a problem (Kurbanova, 2012). The area of pastureland is also in short supply (Halimova, 2012; Strong and Squires, 2012).

## Political Instability and Land Alienation

Political conflict characterizes some areas within the region, for example in Afghanistan, but also in tension within other countries that erupts sometimes in civil disturbance. Population pressure, poverty and resource degradation in adjacent agricultural areas are forcing people to encroach onto the pasture lands that were previously reserved for spring and autumn grazing. The most productive areas of this land were often the key grazing areas used by the transhumant pastoralists and the local villagers as *keystone grazing areas*. The implications of this form of land use change are twofold. First, the quantity of land for existing uses and people diminishes. Second, and more importantly, encroachment is leading to the breakdown of grazing systems that were implemented during the Soviet era has led to accelerated land degradation (Strong and Squires, 2012).

**Box 2. Issues that impact on rangeland management and a failure to benefit from the nation's natural resources**

***Institutional and administrative issues***
Alternative use philosophies and value systems
Understanding existing management systems
Understanding land capabilities
Strengthening social technology at the local level
Ecologically sensitive administrative arrangements
Devolving responsibility
International aid
***Resource rights***
Establishing the legal framework for land use
Facilitating land reform
Empowering disadvantaged minority groups
Re-evaluating the merits of land privatization
Common-property systems
***Providing incentives for sustainable land management***
Recognition of the natural potential of rangeland systems
Macro-economic policy
Trade policy

Tajikistan suffers from a growing divide between the rich and the poor that is causing social conflicts. Government needs to pay more attention to comparative poverty, which is more serious than absolute poverty, because it is a matter of social justice. The rapid increase in the number of landless livestock owners is an emerging challenge for rural development. Land reform has resulted in privatization of pastureland that greatly diminishes the area available for common grazing and this infringes the rights of households (HH) who formerly used it. This has had an adverse impact on the livelihoods of many rural (HH), which have small herds but no land shares in any farm. There are many social, environmental and land tenure factors that make it difficult for land users (Halimova, 2012).

Loss of access to keystone grazing areas (some in the uplands) leads to unaccustomed forms of social stress, makes many traditional resource-husbandry practices irrelevant and forces overgrazing in the area that remain available (Squires, 2012). Opportunities to capture the conservation benefits of the pasturelands are fast disappearing (Lerman, 2012).

## Identifying Outstanding Issues for Tajik Rural Households

Tajikistan is judged to be vulnerable to risks, including livelihood risks and food security risks for its large rural population. Box 2 sets out a list of ongoing issues.

# CONCLUSION

Examination of Tajikistan's rangelands from a social and ecological perspectives can provide new policy insights leading to an ecologically sustainable, economically profitable and socially equitable future for rangeland peoples and the systems on which they depend. Rangelands are characterized by great diversity in climate soils, biota, culture, history, and economic conditions. A shared need for rangeland people is to find ways of coping with episodic events that constrain opportunities for development and that call for flexible and variable management strategies and policy frameworks.

This chapter has attempted to draw together the observations and policy insights that are common to the Tajikistan's, and the broader Central Asian region's, rangelands. Many of the social, ecological and economic problems in these areas have common roots. Often at the periphery of national policy considerations, rangeland people (not only herders and agro-pastoralists but also urban dwellers who are service providers in health, education, commerce, transport and communication) have suffered the side effects that come from activity (including policy formulation) in other areas and sectors. Too often, people who usually do not live in the rangelands had unreal expectations about their ecological and economic potential.

Recognizing these characteristics, the most general recommendations that can be made suggest a need to:

- Begin with a careful assessment of the ecological capacity of rangelands based on local people's aspirations, usually maximizing production per unit area, rather than theoretical bases more suited to commercial ranches in more humid regions where equilibrial systems (see Box 1) prevail.
- Avoid subsidies and price-fixing that mask the capacity of markets to inform land-users about the real economic situation.

- Devolve responsibility for management to local communities: and
- Provide these communities with opportunities to make a significant but sustainable contribution to their economy.

In terms of future research and enquiry, a major seemingly obvious conclusion is that people should be recognized as a valid part of all rangeland ecosystems (Squires 2009) and the region's rangelands should be studied from that perspective. In the past the human dimension has been excluded from most ecological studies of rangeland management. Similarly, the ecological dimension has been either excluded from most social and economic studies of the rangelands, or alternatively, treated in a very simplistic manner.

## REFERENCES

Breman, H., and N. Traoré (Eds.). 1987. "Analyse des Conditions de l'Élevage et de Propositions Politiques et de Programmes," *Mali. Sahel D* (86)302, Club de Sahel/CILSS/OCDE, Paris, 243 p.

Ellis, J. and D. Swift 1988. Stability of African pastoral ecosystems: Alternate paradigms and implications for development *Journal of Range Management* 41(6):450-459.

Halimova, N. 2012. Land tenure reform in Tajikistan: implications for land stewardship and social sustainability: A case study.pp. 305-330 In: Victor Squires (ed) *Rangeland Stewardship in Central Asia: Balancing Livelihoods, Biodiversity conservation and Land protection*. Springer, Dordrecht.

Heshmati, G.A. and V.R. Squires. 2009, New thinking in range ecology, In: *Range and Animal Sciences and Resources Management*, [Ed. Victor R. Squires], in *Encyclopedia of Life Support Systems (EOLSS),* Developed under the Auspices of the UNESCO, EOLSS Publishers, Oxford, UK.

Kurbanova, B. 2012. Constraints and barriers for better land stewardship: Analysis of PRAs in Tajikstan. pp. 129-163 In: Victor Squires (ed) *Rangeland Stewardship in Central Asia: Balancing Livelihoods, Biodiversity conservation and Land protection*. Springer, Dordrecht.

Lane, C. and I. Scoones 1993. *Barabaig natural resource management pp.93-120* In: M.D. Young and O.T Solbrig (Eds). *TheWorld's savannas, economic driving forces, ecological constraints and policy options for sustainable land use*. MAB vol 12. UNESCO, Paris.

Lerman Z, 2012. Rural livelihoods in Tajikistan: What factors and policies influence the income and well-being of rural families? pp. 165-187 In: Victor Squires (ed) *Rangeland Stewardship in Central Asia: Balancing Livelihoods, Biodiversity conservation and Land protection*. Springer, Dordrecht.

Masselli, D. and Imam-ur-Rahim, 2012; Setting the stage: Key features of the present day Central Asian region: an introduction to the wider historical, social, political, economic, cultural and ecological contexts of the Region in a nutshell. pp. 3-30 In: Victor Squires (Ed) *Rangeland Stewardship in Central Asia: Balancing Livelihoods, Biodiversity conservation and Land protection*. Springer, Dordrecht.

Rahminov, R.M. 2012 Evolution of land use in pastoral culture in Central Asia with special reference to Kyrgyzstan.pp. 51-69 In: Victor Squires (ed) *Rangeland Stewardship in*

*Central Asia: Balancing Livelihoods, Biodiversity conservation and Land protection.* Springer, Dordrecht.

Smith, A. 2011. Animal husbandry, nomadic breeding and domestication of animals pp.116-137 In: Victor R. Squires (Ed).*The role of Food, Agriculture, Forestry and Fisheries in Human Nutrition Encyclopedia of Life Support Systems, UNESCO/EOLSS. Oxford UK.*

Squires, V., L.M. Hua., D. Zhang and G. Li (eds) 2009. *Towards Sustainable Use of Rangelands in North West China.* Springer, Dordrecht.

Squires, V.R. and N.M. Safarov 2013. Diversity of plants and animals in mountain ecosystems in Tajikistan. *Journal of Rangeland Science, 4(1):* 43-61.

Strong, P.J.H. and V.R. Squires 2012. Rangeland-based livestock: A vital subsector under threat in Tajikistan pp.213-238 In: Victor Squires (ed) *Rangeland Stewardship in Central Asia: Balancing Livelihoods, Biodiversity conservation and Land protection.* Springer, Dordrecht.

# PART 3. CAPTURING THE CONSERVATION BENEFITS OF BETTER RANGELAND MANAGEMENT

There are 3 chapters in this Part. One of the themes in this book is the importance of ensuring that conservation benefits accrue from managing rangelands in a better way. A focus of the Part is on biodiversity conservation and on remediation of damaged landscapes. The aftermath of a culling program on feral camels from Central Australia provides insights into the unintended consequences of past decisions to re-arrange the fauna in Australia.

Finally, an attempt is made to draw together the divergent strands that comprise the management of rangelands to capture the conservation benefits. A number of important themes are implicit in the title of the book '*Ecology*, '*Management*' and '*Conservation'*. Each of these is important in its own right. Clearly an understanding of the ecology of rangelands underpins the management and provides a direction for efforts to conserve the rangelands for the benefit of current and future generations. There is clear need to integrate ecological and social sciences and develop a classification of rangelands into appropriate regions in order to address inter-regional questions of production potential, conservation value, resilience and socio-economic viability. Rangeland science has commonly only dealt with biophysical constraints. There have been numerous reminders that factors, other than biophysical ones also limit management options and the sustainability of rangeland enterprises, industries and communities. It is now recognized that socio-economic factors are often crucial in determining whether particular technical factors can be employed in practice.

In: Rangeland Ecology, Management ...
Editor: Victor R. Squires
ISBN: 978-1-63482-504-7


*Chapter 8*

# PRELIMINARY INVESTIGATION OF THE DECOMPOSITION OF FERAL CAMEL CARCASSES IN CENTRAL AUSTRALIA

***Benxiang Zeng****
Northern Institute, Charles Darwin University, Alice Springs Campus, Alice Springs, Australia

## ABSTRACT

An integrated national feral camel management plan is being implemented across the camel range in Australia, aiming to control the damage caused by camels. The culling operation is currently the main management approach. Although concerns about the environmental consequences of shot-dead camels decomposing in the bush are often raised, there is a lack of empirical research on this issue.

This study aimed to explore the decomposition process and identify how long the complete decomposition of camel carcasses would take in natural settings in central Australia. The field monitoring was conducted for a consecutive ten months. Monitoring activities included observing, measuring, recording and photo taking. A decomposition model was applied to predict the average decomposition process. SPSS was used for regression analysis and modelling. It was found that camel carcasses would generally take around 2 years to decompose completely in natural settings in central Australia, although the process varied with individual carcasses. The warm and wet weather would accelerate the decomposition process. The preliminary evidences provided in this study would allow a better understanding how culled camels would decompose in the bush. It provides a quantitative basis to explore the environmental consequences of camel management strategy, including a possibility to estimate the carbon emission mitigation through the camel culling program.

**Keywords**: decomposition rate; integrated strategy; feral camel management; monitoring

* Contact: Telephone: +61 8 89595270; E-mail: benxiang.zeng@cdu.edu.au.

# 1. Introduction

There are around 750,000 feral camels in arid and semi-arid rangelands in Australia, mainly distributed on rangelands in Western Australia, South Australia, the Northern Territory and Queensland [Ninti One 2012]. It is estimated that in 2010 the economic losses caused by feral camels were at least AU$10 million [Edwards et al. 2010].

An integrated national feral camel management plan is being implemented across the camel range, aiming to control the damage caused by camels [see http://www.feralcamels.com.au/]. While the culling operation (including aerial and ground shooting) is the main management approach, other approaches such as commercial harvesting are also included in the management strategy. However, they do not make a significant contribution to the camel management regime [Zeng and Gerritsen 2012]. Although concerns about the environmental consequences of shot-dead camels decomposing in the bush are often raised, there is a lack of empirical research on this issue.

This chapter reports the results of monitoring of the decomposition of camels in the bush to provide a better understanding of the environmental impact of the processes involved.

# 2. Decomposition of Dead Animals

Dead animals will decompose in different ways, but they will all undergo the same sequential stages of decomposition. The decomposition stages model identified by Payne [1965] has been commonly used in animal decomposition studies. The six stages are: fresh, bloat, active decay, advanced decay, dry, and remains.

The rate and pattern of decomposition of an animal body is strongly affected by a number of factors, including temperature, availability of oxygen, access by scavengers, wetness, rainfall, body size and weight, surface on which the body rests and the carcass's position and its openness to the elements such as air, water and soil.

There has been no report available about the decomposition of camel carcasses in the wild. However, some studies about the decomposition process of domestic animals and other wildlife have been reviewed.

Using monkey carcasses, Ahmad and Ahmad [2009] documented the decomposition process and found that the biomass loss of carcasses occurred rapidly during the fresh to active decay stage due to the colonization and feeding activity of insects and bacteria. Archer's [2004] study suggested that higher temperatures and rainfall increased both mass loss rates, and decomposition stage progression rates. Parmenter and MacMahon [2009] found that decomposition rates varied significantly between microsites and among seasons, with mass loss amounts linearly correlated with ambient air temperature.

Decomposition studies were conducted using domestic pigs (*Sus scrofa*) to determine differences in rates and patterns of decomposition of carcasses hanging or exposed on the surface of the soil. The result suggested that the rate of biomass removal from the hanging carcass was significantly slower than that observed for the control carcass during the bloat and decay stages of decomposition [Shalaby et al. 2000].

Gonder [2008] monitored the decomposition progress of dead wolves and identified that ambient temperature was one of the important factors to impact on the duration of different stages. It was suggested that during spring to summer it took a much shorter time for carcasses to get through to the dry stage, compared to during autumn to winter time. Temperature was obviously a major impact factor, i.e., the higher the ambient temperature, the more quickly the carcass decayed.

## 3. Methods

An extensive literature review was conducted. The existing studies on animal decomposition were searched with a focus on appropriate methods and specific techniques used to measure and identify the decomposition process of animals, especially in arid and semi-arid areas.

As a main component of this project, on-site monitoring was designed to identify the decomposition process (decomposition stages and associated mass loss) to collect both qualitative information and quantitative data. The collected information included measurements of body size, characteristics of the decomposition process and the length of period (days) of different stages.

Then a decomposition model was applied to predict the average decomposition process. SPSS was used for regression analysis and modelling.

## 4. Research Site and Monitoring Activities

The Curtin Springs Cattle Station was selected as the field site to monitor the process of decomposition of dead camels.

The station is located in central Australia, on the Lasseter Highway, 85 km east of the entrance to the Uluru-Kata Tjuta National Park, 360 km south west of Alice Springs. The station is 4,164 $km^2$ in size. A large number of feral camels are consistently present on the station where they cause damage to infrastructure such as fences and strongly compete with cattle for water and food. The opportunistic culling of camels has been carried out over many years as one of the station's feral animal management measures.

A total of 53 shot-dead camel carcasses were monitored over 10 months. The carcasses were visited once a month and a total of nine site surveys undertaken, from December 2011 to September 2012 (the January 2012 visit was omitted).

On 13 December 2011, the first survey was conducted within the station. Station owners keep records of date of death of all camels shot within the station, which makes it possible to identify the exact dates of death of each camel. Twenty three carcasses of camels were located and monitored. The geographic locations of carcasses were determined with a handheld GPS and the coordinates recorded. During the monitoring, camel sizes (i.e., shoulder height, body length (between legs), thoracic girth/heart girth and abdominal girth) were measured to estimate the weight of the camels when they were shot. Although most of the camels were shot before the day they were first visited, the measuring was still carried out for this purpose. The decomposition stages were observed and described, the percentage of decomposition and left weight of carcass was estimated, and four photographs of each carcass were taken (recording ventral, posterior, dorsal, and anterior views) to visually record the camel's decomposition status. Visual and other observations were noted and included carcass positioning (sitting or lying on their side), desiccation, odour, estimated mass remaining, and any significant disturbance of the carcasses by scavengers. Through this, the decomposition stages were identified.

The surrounding site conditions were observed and recorded, such as saltpan, bushy, grassy or bare land. In addition, general weather data for the Curtin Springs Weather Station was collected through the Bureau of Meteorology website [http://www.bom.gov.au]. This includes daily weather observations (rainfall and temperature) and monthly and historic records.

Further surveys were conducted on 11 February 2012, 15 March 2012, 16 April 2012, 11 May 2012, 13 June 2012, 24 July 2012, 22 August 2012 and 20 September 2012. In these surveys, the monitoring concentrated on the decomposition status of carcasses. The decomposition status was observed and recorded either photographically or by recorded description, including the decomposition stage, mass remaining and other significant attributes (wet, odour, presence of insects, etc.). However, on 13 March 2012 and 16 April 2012, additional 8 and 22 carcasses respectively were added to the monitoring list to make the total number of camel carcasses 53. For these extra carcasses the same monitoring activities were employed as on 13 December 2011.

# 5. Results

## 5.1. Sample

A total of 53 shot-dead camels were monitored. The sample included all age groups and males and females.

Combining the two formulas (Table 1) suggested by Abebe et al. [2002] and Yagil [1994], the weight of camels when they died (i.e., the 'original mass') was estimated.

The original mass of camels ranged from 113 kg to 665 kg, with a mean mass of 385.18 kg and a median mass of 401.43 kg.

The 53 camels were killed in different seasons, between 19 June 2011 (mid winter in central Australia) and 24 March 2012 (late summer). The total decomposition period from the death day until the final monitoring day (20/09/2012) for individual carcasses ranged from 180 to 459 days.

**Table 1. Formulas applied for the original mass estimation**

| | Equation/function | Source |
|---|---|---|
| Formula 1 | $Y_1 = 52 \times SH \times TG \times AG \pm 25$ | Abebe et al. [2002] |
| Formula 2 | $Y_2 = SH \times TG \times AG \times 50$ | Mehari et al. [2007], Yagil [1994] |
| Combined in this paper | $Y = (Y_1 + Y_2)/2$ | |

Note: Y=original mass (kilogram), SH=shoulder height (metre), AG= abdominal girth circumference over the highest part of hump (metre), TG= thoracic/chest/heart girth circumference (metre).

## 5.2. Time for a Complete Decomposition Process

Different carcasses experienced different decomposition processes. During decomposition, carcasses lost more and more of their original mass (Figure 1).

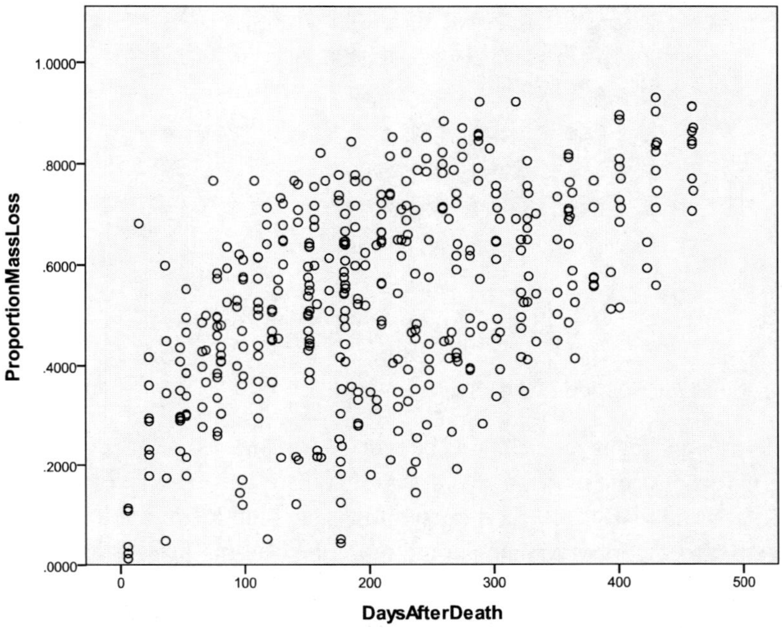


Figure 1. Decomposition process of all carcasses.

**Table 2. Time lengths for different decomposition stages (n=11)**

| Carcass code | Stage 1-3 | Stage 4 | Stage 5 | Stage 1-5 |
|---|---|---|---|---|
| 1 | 66 | 164 | 57 | 287 |
| 2 | 156 | 74 | 57 | 287 |
| 3 | 66 | 164 | 57 | 287 |
| 5 | 208 | 153 | 70 | 431 |
| 7 | 115 | 74 | 70 | 259 |
| 11 | 286 | 74 | 70 | 430 |
| 12 | 327 | 74 | 57 | 458 |
| 14 | 237 | 164 | 29 | 430 |
| 19 | 75 | 164 | 57 | 296 |
| 22 | 145 | 115 | 57 | 317 |
| 23 | 36 | 183 | 70 | 289 |
| **Mean** | **156** | **128** | **59** | **343** |
| **Median** | **145** | **153** | **57** | **296** |

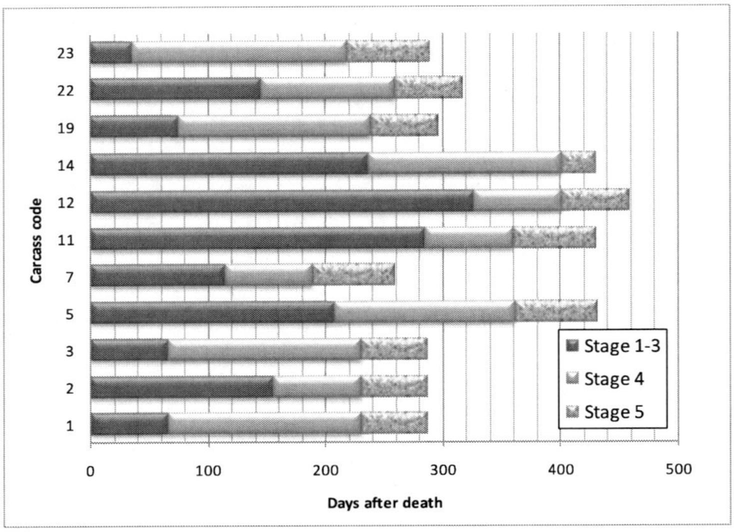


Figure 2. Lengths of DECOMPOSITION Stage 1-5 (n=11).

There is a significant correlation between time (days after death) and degree of decomposition (proportion of mass loss) (r=0.520; sig.=.000; N=395). This suggests that carcasses will continuously decay and decompose and keep losing their mass. The longer the time from death, the less mass remains or the more the original mass is lost.

However, the decomposition process of individual carcasses varied (refer to Appendix III and VI for the detailed data for all carcasses). Based on the observation of 11 carcasses whose decomposition processes moved beyond Stage 5-dry to final Stage 6-remains, it took around 296-343 days on average from culled (Stage 1-fresh) to almost decomposed (Stage 5-dry), the total range varying from 259 days to 458 days (Table 2). It is understood that more time would be required to arrive at a complete decomposition (i.e., Stage 6-remains).

It was observed that both total periods of five stages (Stage 1 to Stage 5) and individual stage periods varied (Figure 2).

## 5.3. Differences in Decomposition Processes of Different Groups

### *5.3.1. Variation between Seasons*

A difference was observed for camels that had died in different seasons. Figure 3 shows the decomposition processes of two groups of carcasses (Group 1: carcasses 1, 2, 3 and 4; Group 2: carcasses 34, 35, 36, 50, 51, 52 and 53). Group 1 were killed on 7 December 2011 (mid- summer), and Group 2 were killed on 24 March 2012 (late summer/early autumn).

It suggests that Group 2 started to decay much more quickly (Stage 1-fresh and Stage 2-bloat only lasted for 6 days or so) and took a much shorter time (66 days or so) to go through to Stage 4-advanced decay, compared with Group 2 (Stage 1-2 lasted for 22 days, and Stage 3-active decay lasted for 180 days or so). This might be explained by the camels killed in summer time having better conditions for decomposition, including higher temperatures combining with more rainfall. This result is consistent with the results suggested by Gonder [2008] in the monitoring of wolf decomposition pattern (refer to Figure 2).

### *5.3.2. Variation in Carcass Positioning*

It was found that camel carcasses in sitting positions decomposed much more slowly and took a much longer time to go through the different decay stages, especially stages 1-3 (Figure 4).

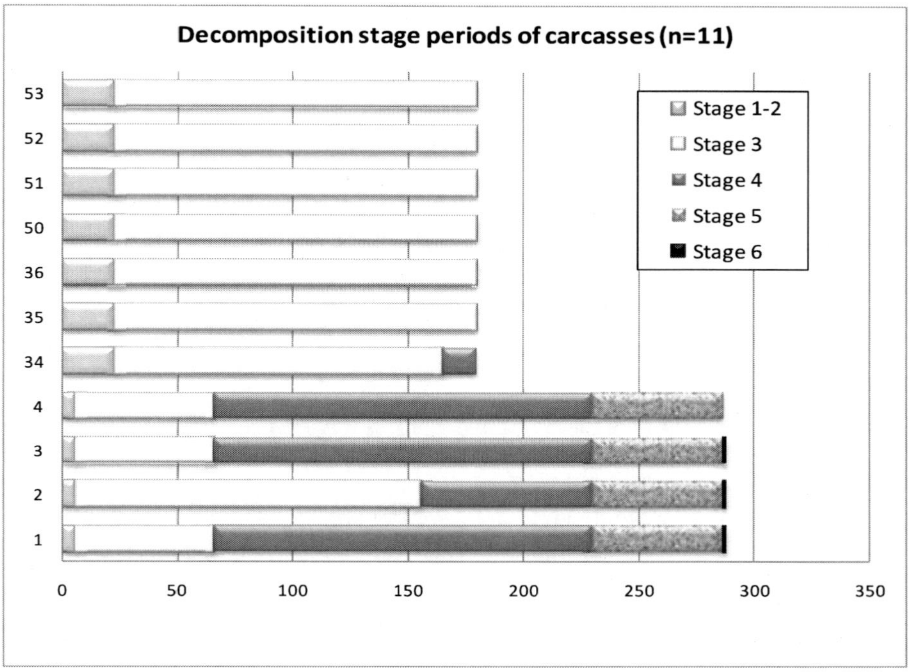


Figure 3. Differences in Stage 1-3 of Decomposition in different seasons (n=11).

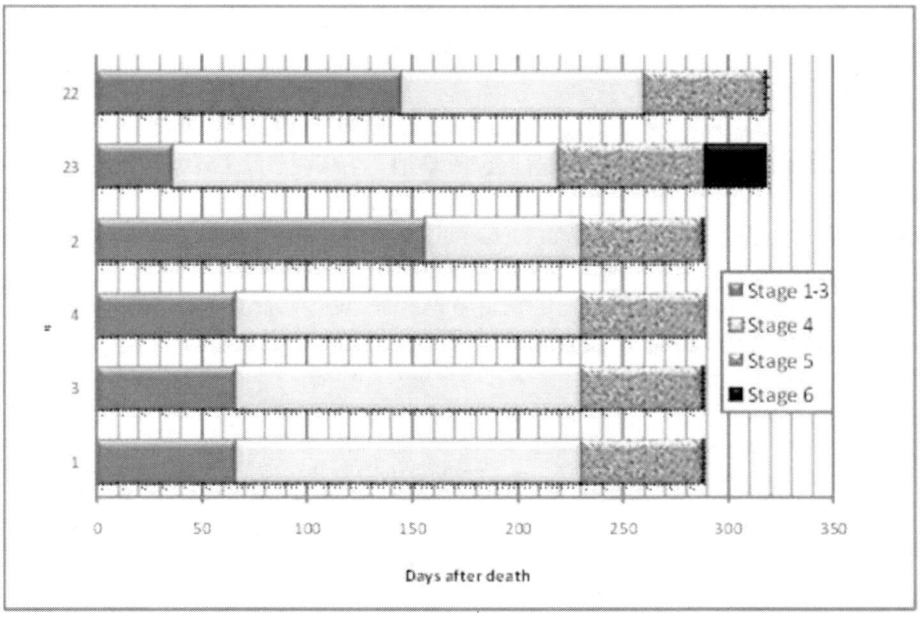


Figure 4. Differences in decomposition stages of carcasses in different positioning.

Figure 5. Camels in a sitting position (above) or in a lying onside position (below).

Camel 2 which was in a sitting position (Figure 5) was killed with Camel 1, 3 and 4 in the same locations and on the same day (December 7, 2011), and Camel 22 which was in a sitting position was killed together with Camel 23 on November 17, 2011. The figure suggests that carcasses in a sitting position took around 150 days to go through to Stage 4-advanced decay, while other carcasses in non-sitting positions only took 35-65 days to go through to the same stage.

In sitting position the softest parts of body (e.g., abdomen) is not exposed and therefore it takes a longer time for microbe and insect biomass to build up. The sitting position also slows scavenging animals such as crows, eagles and dingoes opening up the body, something which positively accelerates the decomposition process.

## 5.4. Modeling the Decomposition Process

A regression analysis was conducted to simulate and model the decomposition process of culled feral camels in Central Australia.

In general, decomposing materials lose mass as they decay and negative exponential models are often used to describe this pattern [Weider and Lang 1982]:

$$\text{Mass}_t = \text{Mass}_0 \times e^{-kt} \qquad (1)$$

where the mass remaining after some time interval (t) is calculated as a function of the original mass and a decay rate coefficient (k). Without knowledge of the underlying mechanisms, the importance of environmental factors can be used empirically to modify this pattern. For example, the rate of mass loss is affected by temperature and moisture conditions.

If: $y = (\text{Mass}_0 - \text{Mass}_t)/\text{Mass}_0$, is the proportion of the mass reduction ( by decomposition) at time t, then: $y = 1 - e^{-kt}$ (2)

Function (2) is used in modelling plant material decay patterns. In this project, I applied it as one of two models to estimate the regression relationship between the decomposition rate and the time from the death for feral camels.

Another possible model for camel carcass decomposition is a logistic growth curve. A sigmoidal curve can be used to model functions that increase gradually at first, more rapidly in the middle growth period, and slowly at the end, levelling off at a maximum value after some period of time. The initial part of the curve is exponential; the rate of growth accelerates as it approaches the midpoint of the curve. At the midpoint, the growth rate begins to decelerate but continues to grow until it reaches an asymptote, C, which is called the "Carrying Capacity" for the environment.

The function is:

$$y = C / (1 + \exp(a + bx)) \quad (3)$$

where, a, C, and b are parameters that shape and scale the function. The value of b is negative.

This type of curve was used as another model in this project. Appling function (3) to this camel carcass decomposition case, C=1 (100% of original mass would be decomposed), -b will be the decomposition co-efficient (as k in function (2)), and x will be time interval t. Thus the function can be presented as:

$$y = 1/(1+e^{a-kt}) \quad (4)$$

Therefore, for the regression analysis, two models were applied:

Model 1: $y = 1 - e^{-kt}$
Model 2: $y = 1/(1+e^{a-kt})$

In the models, $y$ is the proportion of original mass loss, $k$ is the decomposition co-efficient, $t$ is the time interval (days after death), $a$ is constant.

I tried to include all carcasses in the regression analysis, but no models could reasonably simulate the observed decomposition pattern.

However, once all sampled camels were categorised into different groups (based on the discussion in Section 5.3, above), the regression analysis could be established for camels killed either in the spring-summer season (from October to March) or the autumn-winter season (from April to September). As the sitting positioning carcasses had a significantly different pattern from the others (varied but generally slower decay in the early stages), those camels were excluded from the analysis to get better regression relations.

### *5.4.1. Camels Killed in the Spring-Summer Season with Non-Sitting Positioning*

There were 33 sampled camels killed in the spring-summer season, 9 of them were in sitting positioning, and the rest (24) were in non-sitting positioning. Figure 6 shows the decomposition pattern of this group of carcasses.

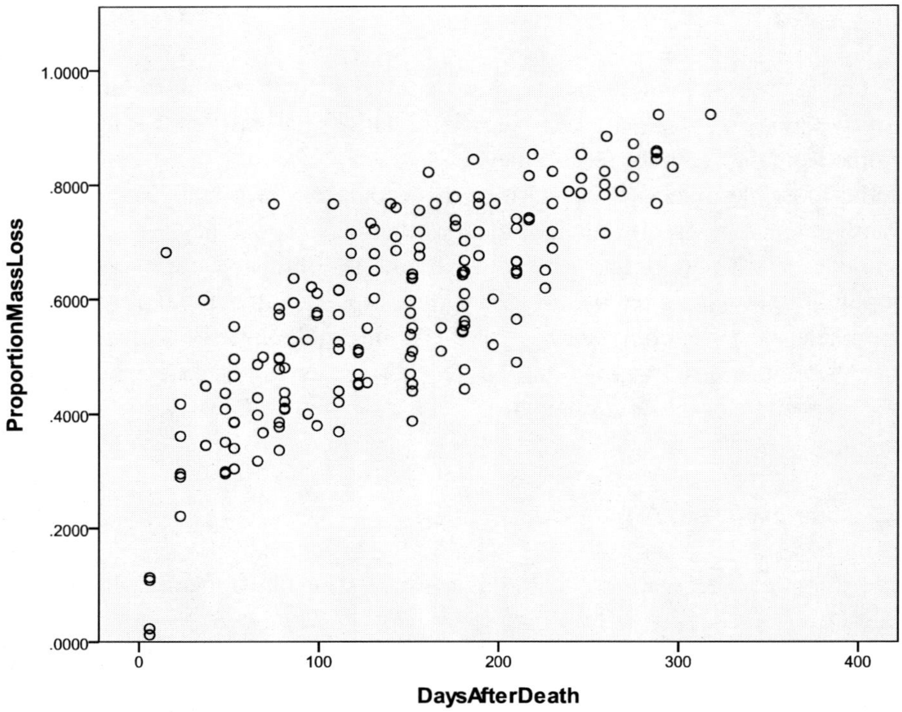


Figure 6. Dot chart showing the decomposition pattern of carcasses in the spring-summer season.

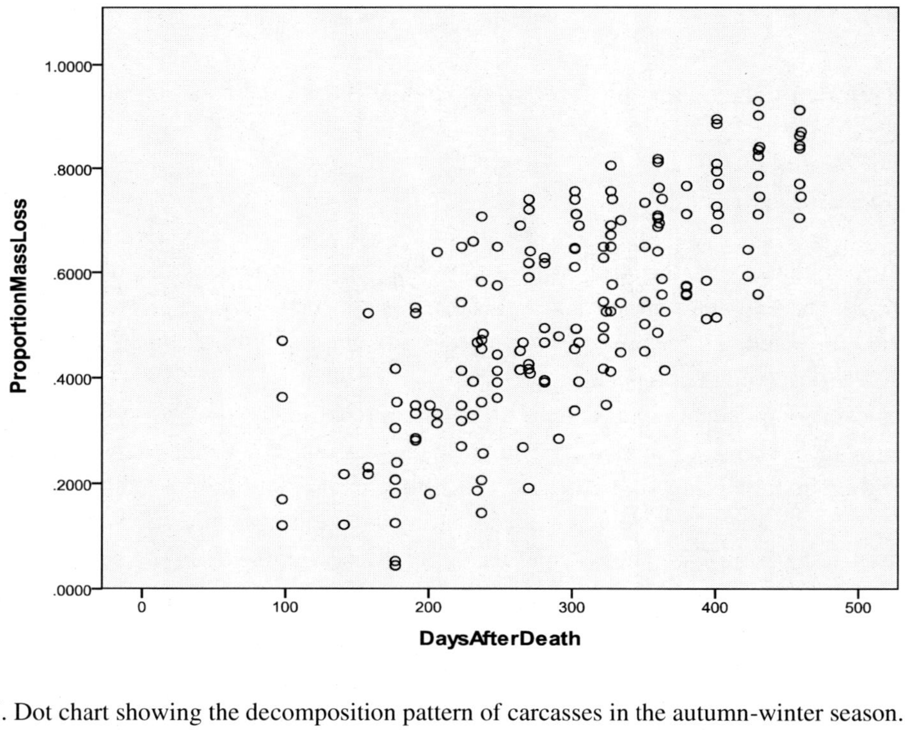


Figure 7. Dot chart showing the decomposition pattern of carcasses in the autumn-winter season.

Appling two models in this group, respectively, two different simulating patterns were returned.

6. $y=1-e^{-0.007t}$,

where $k$ is estimated 0.007 ($R^2$= 0.501).

7. $y=1/(1+e^{0.822-0.008t})$,

where $a$ is estimated 0.822, and $k$ estimated 0.008 ($R^2$=0.631).

Of the two models, model 2 is better, as the $R^2$ (i.e., 0.631) is much greater than that in the model 1 (i.e., 0.501).

### *5.4.2. Camels Killed in the Autumn-Winter Season with Non-Sitting Positioning*

In total there were 20 sampled camels killed in the autumn-winter season, with all of them in non-sitting positioning (Figure 7).

Similarly, the two models were used to conduct regression analysis for camels killed in the autumn-winter season.

After the regression calculation, model 1 was rejected as the Q-squared for this model was too low to suggest this model was accurately simulating the decomposition pattern of this group.

The regression result using model 2 is:

$$y=1/(1+e^{2.102-0.008t}),$$

where $a$ is estimated 2.102, and $k$ estimated 0.008 ($R^2$= 0.570).

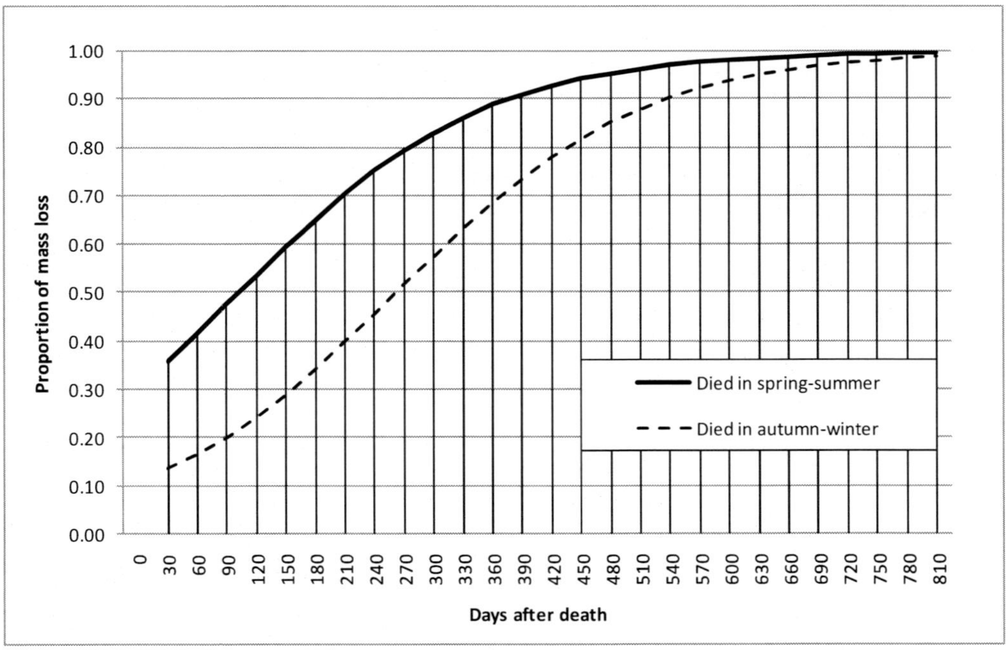


Figure 8. Simulated curves of the mass loss of camel carcass during the decomposition process.

#### *5.4.3. General Estimated Decomposition Pattern of Feral Camels in Central Australia*

Based on the above regression analysis, the general decomposition patterns of feral camels in Central Australia can be simulated as per Figure 8.

These curves provide a general description of the decomposition pattern of feral camels, showing in years how the original mass would be lost during the decomposition process. After 276-436 days (mean: 356 days), the decomposition comes to the final stages (Stage 5/6: Dry/Remains), with only about 20% of total original mass left. After another 100 days or so, the decomposition is nearly complete with only about 10% of original mass left. In sum, it would take around 2 years (or bit longer in some cases) to completely decompose the carcass. Generally speaking, in the spring-winter season, the carcasses would decay much quicker and it would take shorter time to go through to the final stages.

## CONCLUSION

In central Australia camel carcasses take around 2 years to be completely decomposed, although the process varies with individual carcasses. Camels culled in the spring-summer season decay quicker than those culled in the autumn-winter season. This is associated with the combination of high temperature and more rainfall in the spring-summer season in central Australia. The positioning of carcasses also affects their decomposition rates.

This project provides a preliminary study of the decomposition process of feral camels on central Australian rangelands. The evidences provided here would allow a better understanding how culled camels would decompose in the bush. It provides a quantitative basis to explore the environmental consequences of camel management strategy, including a possibility to estimate the carbon emission mitigation through the camel culling program.

This research has limitations. The results could be improved and made more accurate by modifying the research methods. Quantitative data—such as original mass and mass loss—could be improved by attending sample camels immediately after their death, intensifying the monitoring interval (e.g., more frequent site visits), increasing the number of monitoring sites and including different regions, extending the monitoring period (e.g., to 24 months), as well as increasing the number of sampled carcasses. Qualitative data, such as identification of decomposition stages, could be improved by a multi-observation approach, where the each site observation is conducted by more than one researcher/observer. Such observation results could be triangulated to achieve more accurate description and identification.

## ACKNOWLEDGMENTS

This chapter is based on the project 'Engaging Aborigines to monitor feral camels' impacts on carbon emission' funded by the EnvironmeNT Grants Program. Special thanks should go to: Ashley and Lyndee Severin, owners and managers of the Curtin Springs Cattle Station, for their kind cooperation and support during the on-site monitoring; Glenn Edwards, Kym Schwartzkopff, Jayne Brim-Box, Jude Prichard and Peter Barker, for their information sharing and kind support in the early stage of the project; Stephen Hamilton and Rolf Gerritsen, for their valuable comments and editing of previous version of this chapter.

## REFERENCES

Abebe, W., Getinet, A.M. and Mekonnen, H.M. (2002). Study on live weight, carcass weight and dressing percentage of Issa camels in Ethiopia. *Revue Méd. Vét., 153 (11),* 713-716.

Ahmad, A., and Ahmad, A. H. (2009). A preliminary study on the decomposition and dipteran associated with exposed carcasses in an oil palm plantation in Bandar Baharu, Kedah, Malaysia. *Tropical Biomedicine, 26(1),* 1-10.

Archer, M. S. (2004). Rainfall and temperature effects on the decomposition rate of exposed neonatal remains. *Science & Justice, 44(1),* 35-41.

Edwards, G. P., Zeng, B., Saalfeld, W. K. and Vaarzon-Morel, P. (2010). Evaluation of the impacts of feral camels. *The Rangeland Journal, 32 (1),* 43-54.

Gonder, C. (2008). *Wildlife Decomposition in West Central Montana: A Preliminary Study Conducted to Provide Field Investigation Material and Training for Wildlife Officers.* Masters Thesis, University of Montana, Missoula, MT, USA.

Mehari, Y., Mekuriaw, Z. and Gebru, G. (2007). Potentials of camel production in Babilie and Kebribeyah woredas of the Jijiga Zone, Somali Region, Ethiopia. *Livestock Research for Rural Development, 19 (4). Article #58.* Retrieved December 14, 2012, from http://www.lrrd.org/lrrd19/4/meha19058.htm.

Ninti One (2012). Media Release (14 September 2012). Australia's feral camel population declining. Available online: *http://www.feralcamels.com.au/ sites/default/files/upload/ mr_100_000_feral_camels_removed_14sep12.pdf,* accessed on 25 September 2012.

Parmenter, R. R., and MacMahon, J. A. (2009). Carrion decomposition and nutrient cycling in a semiarid shrub-steppe ecosystem. *Ecological Monographs, 79(4),* 637-661.

Payne, J.A. (1965). A summer carrion study of the baby pig Sus scrofa Linnaeus. *Ecology, 46,* 592–602.

Shalaby, O. A., deCarvalho, L. M. L., and Goff, M. L. (2000). Comparison of patterns of decomposition in a hanging carcass and a carcass in contact with soil in a xerophytic habitat on the Island of Oahu, Hawaii. *Journal of Forensic Sciences, 45(6),* 1267-1273.

Wieder, R.K. and Lang, G.E. (1982). A critique of the analytical methods used in examining decomposition data obtained from litter bags. *Ecology, 63,* 1636-1642.

Yagil, R. (1994). *The camel in today's world, a hand book on camel management.* Germany-Israel Fund for Research and International Development and Deutsche Welthungerhilfe, Bonn, Germany.

Zeng, B. and Gerritsen, R. (2012). Inadequate contribution of commercial harvest to the management of feral camels in Australia. *Journal of Environmental Planning and Management,* DOI:10.1080/ 09640568.2012.716367.

In: Rangeland Ecology, Management ... ISBN: 978-1-63482-504-7
Editor: Victor R. Squires 

*Chapter 9*

# RESILIENCE MANAGEMENT OF ALPINE RANGELAND ECOSYSTEMS UNDER CLIMATE CHANGE IN NORTH-WEST CHINA

***Victor R. Squires, Hua Limin and Zhang Degang***
Rangeland Resources and Ecology Group,
Gansu Agricultural University, Lanzhou, China

## ABSTRACT

Alpine meadow ecosystems, a very special kind of rangeland, occupy a large area in the mountainous regions of north-west China. The principal areas are the Qinghai-Tibet Plateau, the northern parts of the Qilian Mountains in Gansu and the Altai Shan, Tian Shan and Kunlun Shan in Xinjiang. Altitudes in these alpine areas range from 2500- over 6000 m above sea level (a.s.l.). Temperatures are rising four times faster than elsewhere in China, and the Tibetan glaciers are retreating at a higher speed than in any other part of the world. In the short term, this will cause lakes to expand and bring floods and mudflows. In the long run, the glaciers are vital lifelines for Asian rivers, including the Mekong, Indus, Brahmaputra and the Ganges. Once they vanish, water supplies in those regions will be in peril.

Alpine areas are particularly vulnerable to climate change because of the effect of altitudinal compression that allows major changes over short distances. Global climate change, specifically increases in temperature, and carbon dioxide ($CO_2$), combined with anthropogenic threats is already threatening the resilience of alpine ecosystems and various vulnerable species - both plant and animal. The effect of increases in $CO_2$ will vary according to species and local environmental condition. On the positive side, scientists predict that short-term $CO_2$ enrichment will lead to increases in photosynthesis, growth, total biomass, root/shoot ratio, and tissue carbon/nitrogen ratio in alpine plants but some species may be more favored than others and this could accelerate the already existing shifts in botanical composition.

# 1. Introduction

North-west China is dominated by the immense Qinghai-Tibet Plateau and the adjacent front ranges (Qilian Shan (in Gansu) and the Kunlun Shan and Tian Shan and Altai Shan in Xinjiang (Figure 1).

Figure 1. Map showing relative position of Tibet, Gansu, Sichuan, Yunnan and Xinjiang in west China in relation to neighbours like India, Bhutan, Nepal and Burma (Myanmar).

The Tibetan Plateau is surrounded by massive mountain ranges. The plateau is bordered to the south by the Himalayan range, to the north by the Kunlun Range which separates it from the Tarim Basin, and to the northeast by the Qilian Range which separates the plateau from the Hexi Corridor and the Gobi Desert. To the east and southeast the plateau gives way to the mountainous headwaters of several of Asia's most important rivers (Figure 2) including the Yangtze, Ganges and the Mekong.

The Tibetan Plateau is bounded on the north by a broad escarpment where the altitude drops from around 5,000 metres to 1,500 metres) in less than 150 kilometres. Along the escarpment is a range of mountains. The easternmost range is the Qilian Shan. North of the mountains runs the Gansu or Hexi corridor which was the main silk-road route from China proper to the west. These regions include some of the world's highest mountains, and the most extensive glacier fields. The Tibetan Plateau contains the world's third-largest store of ice. This is significant for several reasons (i) In the long run, the glaciers are vital lifelines for Asian rivers, including the Indus and the Ganges (Figure 2). Once they vanish, water supplies in those regions will be in peril. (ii) Cold water can absorb more $CO_2$ than warm water and as $CO_2$ concentrations rise, carbon sequestration becomes important (see below).

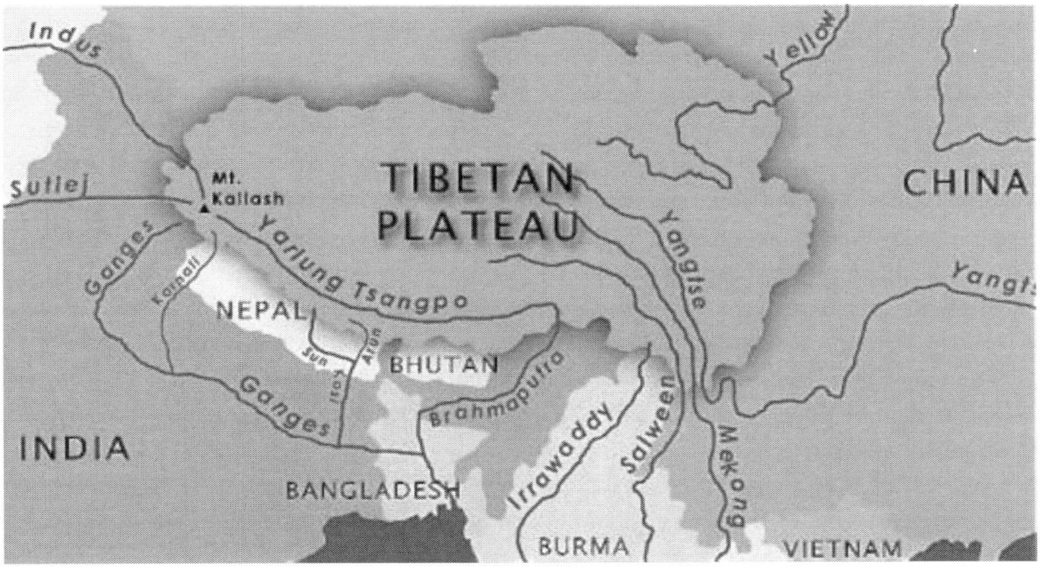


Figure 2. Many of Asia's major rivers arise in the alpine regions of the Tibet Plateau.

## 2. Vegetation in the Alpine Regions

The Tibetan Plateau, treeless except in the south-eastern river valleys, supports a range of alpine vegetation types that includes meadow, steppe, cold desert and sub-nival cushion plant communities at elevations ranging from 3,500 to nearly 6,000 m. Dry, cold, and expansive, the Tibetan Plateau possesses an alpine landscape of complex zonation with a general trend from moist alpine scrub to steppe vegetation to high, cold desert along a transect from southeast to northwest.

The plateau is a high-altitude arid steppe interspersed with mountain ranges and large brackish lakes. The southern and eastern edges of the steppe have grasslands which can sustainably support populations of nomadic herdsmen, although frost occurs for six months of the year. Permafrost occurs over extensive parts of the plateau. Annual precipitation ranges from 100 to 300 mm and falls mainly as hailstorms. The plateau becomes progressively higher, colder and drier, until reaching the remote regions in the north-western part of the plateau. Here the average altitude exceeds 5,000 m a.s.l. and winter temperatures can drop to −40°C. As a result of this extremely inhospitable environment, the Changthang region (together with the adjoining Kekexili region) is the least populous region in Asia.

There are extensive areas of alpine meadows, principally in the altitudinal belt between 2500 and 5000 meters above sea level (a.s.l.). Above this upper altitudinal limit, the vegetation gives way to other ecosystems. Alpine shrublands, characterized by *Rhododendron*s, predominate at lower elevations, close to the treeline. The rhododendron flora of the ecoregion is quite varied, with species composition changing as one moves from west to east. Within the alpine region, moisture determines whether a region supports meadow, steppe or alpine desert vegetation. As precipitation decreases north-westward, vegetation changes from dense scrub to meadow to steppe to desert. Above the shrublands are alpine meadows which support a variety of herbaceous plants, including species of

*Rhododendron, Sedum, Draba, Gentiana, Impatiens, Kobresia, Leontopodium, Meconopsis, Pedicularis, Potentilla, Primula, Saussurea, Saxifraga, Alchemilla, Androsace, Anemone, Diapensia,* In the spring and summer, the alpine meadows are covered with brightly colored flowers. On the upper slopes, low grasses and cushion plants grow among the boulders and scree.

The extensive highland area of the Altai Shan landscape in northern Xinjiang, for example, with cold temperatures and short growing season, supports alpine meadows and tundra, a vegetation type that constitutes a primary component.

High plateaus extend for hundreds of square kilometres ($km^2$) with lichens and mosses as the principal ground cover. Below 2,400 m, subalpine conifer forests cover a large area.

Dominant tree species here are Siberian larch (*Larix sibirica*) and Siberian fir (*Abies siberica*). Stands of these trees are part of a mosaic landscape that also includes meadow with an alpine component.

Thus, at its lower margin, the distinction between the ecoregions Altai Alpine Meadow and Tundra and Altai Montane Forest and Forest Steppe tends to blur.

## 3. Important Role of Alpine Meadows

Alpine meadows, which lie above the tree line but below the snow line, function as 'sponges' by absorbing melting snow and acting as natural water towers. Tibetan people have had a hand in creating and maintaining meadows over the past 2000 years through land clearing, burning and grazing. Meadows served as sites for gathering food and medicinal plants and over 40 % of contemporary species are ethno-botanically useful. Remote sensing analyzes indicate a substantial (69.6 %) decline in meadow area between 1974 and 2004 (a trend that continues). Many Tibetans report a loss of their "true history" and connections to the past associated with loss of meadows (Urgenson et al., 2014). Conservation policies intended to preserve biodiversity are unintentionally contributing to the loss of these ecologically and culturally significant meadow habitats so interventions need to be thoughtfully implemented.

They typically have high herbaceous species richness and are filled with plants well-adapted to harsh climates. For example, nearly a third of plant biodiversity is found above 4,000 metres, and some plants live at above 5,500 metres. But although alpine plants are well-adapted to cold climates, they also tend to be highly sensitive to climatic changes (Squires and Safarov, 2013). Many of the alpine region are deserts. The alpine deserts of northern Tibet include the lofty plateau that lies between the Karakoram and Kunlun mountain regions and extends eastward along the northern margin of the plateau. The sandy grasslands, a feature of deserts in the alpine areas of the Qinghai-Tibetan Plateau, are increasingly influenced by desertification due to both climate warming and human activities such as livestock over-grazing and land use change (Wu and Du, 2007). Desertification of sandy grasslands impacts aboveground productivity, vegetation biodiversity, and organic carbon (SOC), N, and P accumulation and storage in topsoil (Li et al., 2009). Restoration of these degraded sandy grasslands is an important ecological issue, especially in the alpine areas of harsh climate and environment. Recently, alpine grasslands on the Tibetan Plateau have suffered from severe degradation, including desertification (Wu and Du, 2007). This has decreased vegetation condition and ecological quality with severe pest damage, species loss, grassland desertification, and soil erosion and degradation (Hao, 2008). Restoration and rehabilitation of degraded terrestrial ecosystems offer potentially good economic prospects (Dong, Li and Li 2010, Shang, et al., 2014).

## 4. Potential Impacts of Climate Change on Alpine Ecosystems

Global climate change, specifically increases in temperature, $CO_2$, storms, and precipitation, combined with anthropogenic threats will threaten the resilience of ecosystems and species in north-west China.

### 4.1. Effects of Changes in Temperature

Since 1880, the earth has warmed 0.6-0.8°C (Houghton et al. 2001) and it is projected to warm between 2-4°C by 2100 mostly due to human activity (IPCC 2007). The comparison between the historical (1850) and the recent (2007–2010) ambient air temperature data revealed a warming trend in the study area during the last century. The mean temperature of the warmest and the coldest months increased by 0.76±0.25°C and 3.65±2°C, respectively. The changes in daily maximum and minimum temperatures between the two time periods resulted in a narrowing of diurnal temperature range (DTR) by an average of 3°C. The difference between the two time periods was most significant for the higher elevational band of 4400–4600 m compared to the lower bands. The lapse rate appears to have reduced from 0.82°C/100 m in 1850 to 0.42°C/100 m in 2007–2010. Notably, the results showed more warming in winters than in summers, with the differences in mean temperatures of the coldest and the warmest months during the century being 3.75°C and 0.5°C, respectively. At higher elevations (5300–5500 m) the changes in the mean temperatures of the coldest and the warmest months during the last ~150 years were 5.6°C and 0.8°C, respectively. The lower altitudinal band (3200–3500 m) recorded a change of 1.6°C and 1.0°C in the mean temperatures of the coldest and the warmest months, respectively.

### 4.2. Effects of Changes in $CO_2$

Atmospheric $CO_2$ has increased from 280 parts per million by volume (ppmv) in 1880 to nearly 379 ppmv in 2005, and about 30% of all atmospheric $CO_2$ resulting from fossil fuels has been taken up by the ocean (IPCC 2007). The implication of high $CO_2$ concentrations for alpine plants is clear. Scientists predict that short-term $CO_2$ enrichment will lead to increases in photosynthesis, growth, total biomass, root/shoot ratio, and tissue carbon/nitrogen ratio in alpine plants but some species may be more favored than others and this could accelerate the already existing shifts in botanical composition (see below). Another important factor is that cold water can absorb more $CO_2$ than warm water and as $CO_2$ concentrations rise, carbon sequestration becomes important to mitigating the effect of global warming.

### 4.3. Effects of Changes in Precipitation

By 2050, a 25% increase in precipitation rates is predicted due to global warming. However, the precipitation will be unevenly distributed with either increases or decreases in different locations. Decreased precipitation results in lowered productivity, growth, and seedling survival, and may change species composition favoring more drought tolerant species. Increased precipitation is likely to increase run-off, erosion, and silt deposition and increase turbidity and reduce primary productivity in lakes and reservoirs.

### 4.4. Effects of Changes in High Intensity Rainstorms

Climate change will also cause storms to increase in intensity (IPCC 2007) and possibly to increase their frequency further stressing alpine ecosystems. Higher rainfall intensity can accelerate soil erosion and bring about flash flooding and landslides.

## 5. Impact of Climate Change on Alpine Regions

Temperatures are rising four times faster in NW China's alpine regions than elsewhere in China, and the Tibetan glaciers are retreating at a higher speed than in any other part of the world (Lu, Wang and Wu, 2009). In the Qilian Shan there are 2815 glaciers covering 1930 $km^2$. The glaciers are melting and shrinking. In the short term, this will cause lakes to expand and bring floods and mudflows. In the long run, the glaciers that are vital lifelines for Asian rivers, including the Mekong, Indus, Brahmaputra and the Ganges will have diminished water yield. Once the glaciers vanish, water supplies in those regions will be in serious peril and lives of hundreds of millions of people will be in jeopardy.

Alpine areas are particularly vulnerable to climate change. Global climate change, specifically increases in temperature, and carbon dioxide ($CO_2$), combined with anthropogenic threats is already threatening the resilience of alpine meadow ecosystems and various vulnerable species - both plant and animal. The effect of increases in $CO_2$ will vary according to species and local environmental condition, but scientists predict that short-term $CO_2$ enrichment will lead to increases in photosynthesis, growth, total biomass, root/shoot ratio, and tissue carbon/nitrogen ratio but some species may be more favored than others and this could accelerate the already existing shifts in botanical composition (see below).

### 5.1. Effects of Changes in Precipitation

By 2050, a 25% increase in precipitation rates is predicted due to global warming. However, the precipitation will be unevenly distributed with either increases or decreases in different locations (Houghton et al. 2001). An analysis of precipitation and temperature changes in several alpine meadows in NW China show that the trend is not uniform (Wu and Li, 2009) Lu, Wang and Wu, 2009, analyzed temperature and precipitation for eight sites in the N and NW of China and showed that temperatures have risen and in most cases precipitation is higher. Rising temperatures cause glaciers to melt and in Xinjiang and Gansu, two of China's NW regions, glacial retreat is quite serious. Initially, the increased melt water may be welcome but it can lead to flooding and bring about landslides. Where there is no calamity, local herders welcome the release of more potential grazing land from the retreating ice (Squires et al., 2009).

Decreased precipitation results in lowered meadow productivity, growth, and seedling survival, and may change species composition favoring more drought tolerant species. It is also likely to result in decreased meadow area and in diversity in the longer term, not only of plants but some ground burrowing vertebrates (plateau pika *Ochotona curzoniae*) and a range of invertebrates.

Decreased precipitation results in lowered meadow productivity, growth, and seedling survival, and may change species composition favoring more drought tolerant species. It is also likely to result in decreased meadow area and in diversity in the longer term, not only of plants but some ground burrowing vertebrates and range of invertebrates. Another outcome is the changing botanical composition of the alpine area as more drought-tolerant species, particularly shrubs, can invade and become dominant. Emerging research in Asia appears to support the emerging hypothesis that climate change is a key factor contributing to shrub encroachment in alpine meadows, and that further warming may threaten the survival of rare alpine plants (Telwala et al., (2013) found that 87% of 124 endemic plants surveyed in the Himalayas were migrating up mountains, partly in response to a warming climate. Their study, which compared contemporary data with 19th century records, warned that continued warming there may lead to "exacerbated species extinction" in upper alpine regions. Climate change and competition from drought-tolerant species that live at lower altitudes is threatening dozens of alpine plants, in part because the potential to escape to suitable cold habitats is highly limited. The loss of more alpine plants would be a great loss to the whole world because these species have not been scientifically investigated in detail.

Numerous studies have shown that climate warming has induced species' geographic range shifts and occasional extinctions. The warming induced upward shifts of species occur along latitudinal and altitudinal clines. A number of studies have suggested alpine plants to have evolved adaptations in response to the long-term climate change from the Pleistocene to late Holocene epochs.

Low temperature is considered as the main controlling factor of alpine plant life in a mountain ecosystem. Therefore, montane floras, in particular, are highly sensitive to climate change since their distribution is mainly controlled by climate related ecological factors that dominate at high altitudes. Mountains are thus, regarded as particularly suitable observation sites for tracing climate-change impacts. The Himalayan gradient represents the longest bioclimatic elevational gradient and constitute the highest elevational limits of vascular plant species (6400 m) on Earth; these montane ecosystems are undergoing rapid warming, increase in precipitation, and extensions of the growing season. Alpine plant diversity of the Himalaya is higher than the global average and significantly higher plant diversity and richness is found between 4200 and 4500 m elevations compared to similar elevations of any other mountain range.

### 5.2. Effect of Climate Change on Species' Ranges and Richness

The observed upward shifts in the species' range margins were categorized into four response types: (i) species in which both upper and lower margins shifted (52%), (ii) species in which only the lower margin shifted (19%), (iii) species in which only the upper margin shifted (20%), and (iv) species in which neither upper nor lower range margins shifted (9%). The species of the lower elevational band of alpine region (4000–4500 m) showed more pronounced upward shifts compared to the species of higher elevational bands (>4700 m). Notably, all the plant species with the largest range shift of 600–800 m were from the lower elevational band of the alpine zone (4000–4500 m). More than 50% of species showed an overall range expansion of 100–300 m in extent, while 62% species showed range contraction of 100–300 m.

Adjustments to elevational ranges or migration of plant species into alpine areas occurs in order to occupy more suitable habitats and retreat from less viable areas under the stressor of climatic warming. These changes are reflected either in the upward shifts of the range margins or increase in species richness at mountain peaks. In this study there was a pronounced upward shifts in species' ranges across the endemic species and plant families studied. The critical determinant of elevational shift is the recorded increase in the minimum temperatures of about twice that of maximum temperatures in the high mountains.

The critical determinant of elevational shift is the recorded increase in the minimum temperatures of about twice that of maximum temperatures in the high mountains. Although sizeable habitable areas exists in the high-altitude (above 5,500 m), it is unclear whether many Himalayan alpine plant species will be able to shift northwards at sufficient pace during the 21$^{st}$ century. Indeed, the smaller shifts observed in the upper range margins and the declining species richness (especially at higher altitudes, is a serious challenge that requires continuous monitoring. With no more habitats available further northward, these species will likely suffer from compression of range extents. If the climate warming trend continues these species may become locally extinct. The wide taxonomic range of the species exhibiting altitudinal shifts suggests that there are no phylogenetic confounding factors or clustering, but instead a common environmental variable contributing to the migration. Temperature being the main controlling factor of alpine plant life, investigations suggest that the species range shifts across phylogenetic affiliations are a direct result of warming.

Modelling studies have shown that species at lower elevations (sub-alpine) have a wider range tolerance and more availability of suitable habitats. These observations are consistent with findings that the elevational range shift of lower band alpine species (4000–4500 m) are more pronounced compared to species of upper elevational bands. It is also likely that the species of lower alpine zone have better adaptability to colonize new habitats as opposed to the species at higher elevational zone. As a result, the failure to adapt at higher elevational zones may result in exacerbated species extinction.

## 6. Managing for Climate Change

We need to lend whatever support we can to China's climate change mitigation effort until it yields positive results, while doing all we can to ensure the survival of ecosystems and species in the interim. If we were to be able to address climate change simply by coming up with management strategies that promoted adaptation in natural systems and populations, we would be lucky indeed. The challenge is larger than that and confounded by an overlay of anthropogenic stressors that further weaken ecosystems, inhibit their ability to recover or adapt, and hasten phase shifts to irretrievably altered states or local extinction of populations. Managing alpine meadows for resilience offers us a potentially effective means to address the present climate and anthropogenic challenges, as well as the uncertainties of the future. We need to recognize a fundamental truth: the only long-term solution for addressing climate change is to roll back greenhouse gas emissions to tolerable levels. Designing and managing protected areas (PAs) for resilience offers us a potentially effective means to address the present climate and anthropogenic challenges, as well as the uncertainties of the future.

## 6.1. Resilience Management

Sustainability involves maintaining the functionality of a system when it is perturbed, or maintaining the elements needed to renew or reorganize if a large perturbation radically alters structure and function (Squires, Chapter 1, this volume). The ability to do this is termed "resilience. A fundamental difficulty in managing social-ecological systems (SESs) for long-term, sustainable outcomes is that their great complexity makes it difficult to forecast the future in any meaningful way. Not only are forecasts uncertain, the usual statistical approaches will likely underestimate the uncertainties.

According to Walker et al., (2002) 'even the uncertainties are uncertain'. They say that 'there are several reasons why uncertainties are large and difficult to characterize:

- Key drivers, such as climate and technological change, are unpredictable. Many change non-linearly.
- Human action in response to forecasts is reflexive. If important ecological or economic predictions are taken seriously, people will react in ways that will change the future, and perhaps cause the predictions to be incorrect.
- The system may change faster than the forecasting models can be recalibrated, particularly during turbulent periods of transition, so forecasts are most unreliable in precisely the situations where they are most wanted.

These aspects of uncertainty limit the usefulness of forecasting methods for the scientific study and management of regions in transition. Given these limits to understanding, we must focus on learning to live within systems, rather than "control" them.

Because of our limited understanding of the impacts of climate change on meadow ecosystems in NW China and how to address them, it is difficult to propose with confidence specific management strategies for climate change adaptation. Consequently, it becomes increasingly critical to maximize the resilience or capacity of the ecosystem to cope with changes generally. Management for resilience is thus not only a general strategy for conservation, but an important part of responding to the impending threat and uncertainties of climate change. Understanding ecosystem dynamics and the factors that help them to resist or recover from perturbations can help us to develop management responses that are designed specifically to address climate change and build resilience to its impacts. Management for resilience is thus not only a general strategy for conservation, but an important part of responding to the impending threat and uncertainties of climate change.

Establishment of Protection Areas (PAs) have been identified as one of the most effective tools for conserving ecosystems. However, PA managers will need to incorporate climate change along with increasing human impacts into their conservation strategies, or PAs may not be able to safeguard biodiversity and resource values effectively. Networks of PAs, provide a critical means to protect alpine systems from human stresses and, if designed for resilience, from climate change stressors as well. The key to addressing climate change impacts is the acknowledgement of the threat it poses and being proactive in developing and applying adaptation strategies and building sufficient flexibility into management systems to enable adaptive responses.

## 6.2. What Can PA Managers Do?

While there is little that PA managers can do to control climate change, the "Resilience Model" developed by the Nature Conservancy of UK, provides a simple tool that is being increasingly used to build resilience into PA networks. The six strategies of the model and their application to PA network design are described below.

Strategy 1: *Apply risk-spreading strategies to address the uncertainties of climate change:* Protect multiple examples of a full range of habitat types, seeking to represent the area's total biodiversity. Replication within each habitat type reduces the chance of any one type being completely compromised by an unmanageable impact. To fully represent regional biodiversity within PAs, aim to protect 30% of the complete range of habitat types. Maintaining biodiversity can enhance resilience if sufficient functional redundancy exists to compensate for species/habitat loss. This approach also has the advantage of protecting essential habitat. Managers will need to develop a classification scheme of habitat types and major zones, and categorize these by their biodiversity, condition, level of threat, resilience, and ecosystem services.

Strategy 2: *Protect critical areas that are resilient to climate change and that serve as refuges to reseed affected areas:* Identify and fully protect communities or habitats that have resisted or recovered rapidly from climate change impacts and that can serve as refuges and provide seed to repopulate and enhance recovery of areas damaged by catastrophic events. Areas that demonstrate persistence over time are important sites to protect. They are refuges for biodiversity. Refugia should be established and must be large enough to support high species richness. Large areas of suitable habitat maintain high species richness as well as high genetic diversity because they may support larger populations that produce more offspring.

Strategy 3: *Understand and preserve connectivity between habitats to enhance mutual replenishment and recovery and to maintain functional linkages among associated habitats:* Identify patterns of connectivity among source and sink areas and among associated habitats. Connectivity describes the natural linkages among habitats and ecosystems that result from the dispersal of organisms. The strength of connectivity depends on the abundance and fecundity of source populations, the longevity and movement patterns of adults. Connectivity is thus a key driver of the level and reliability of biodiversity replenishment in habitats damaged by natural or human-related agents, including climate change. Managers should also endeavour to maintain the connectivity between upland water catchments and downstream areas.

Strategy 4: *Manage ecosystems for both health and resilience, and monitor multiple indicators of the effectiveness of these actions as the basis for adaptive management:* Control human threats to PAs and their conservation targets, monitor the response of species, communities, and ecosystems to climate change, and manage adaptively to compensate for changes in species ranges and environmental conditions and accommodate new science-based strategies as knowledge increases. Because unstressed ecosystems will be better able to resist shocks and adapt to change, PA managers will need to reduce anthropogenic threats to them and thereby strengthen their resilience to climate change.

Strategy 5: *Restore degraded critical areas that have high survival prospects* Areas that are currently degraded but that meet resilience criteria should be restored. Successful techniques for this have been developed (Shang et al., 2008, Dong, et al., 2008, Wu et al., 2009b). Banning of grazing is a widely used means of restoring degraded rangeland in China (Zhang et al., 2015). However, little is known about the time required to restore degraded alpine meadows through the use of a grazing ban. Height and cover of individual plant species and total cover, herbage mass and root mass of each plot of a grazing ban and communal free-grazing alpine meadows in Northern Tibet were examined. Soil samples were also collected and total organic carbon, microbial biomass carbon, total phosphorus and ammonium nitrogen contents were measured.

The results showed that both the level of plant species diversity and herbage mass were higher in areas of a grazing ban than in communal free-grazing land without a grazing ban. No significant differences in root mass and soil nutrient contents were observed. There was a higher plant species diversity and herbage mass in the early years of a grazing ban but there was a subsequent decline. It is suggested that some grazing after a grazing ban may be necessary according to the dynamics of ecosystem responses with time.

Strategy 6: *Manage Functional Groups.* A functional group is a collection of species that perform a similar function, irrespective of their taxonomic affinities. Functional groups are collections organisms based on morphological, physiological, behavioral, biochemical, or environmental response. Some areas have more species in each functional group; thus they have functional redundancy. Communities with functional redundancy may have a better chance of recovery if one species is lost from a functional group.

## 7. Implement Monitoring and Adaptive Management

Because of the anticipated increasing level of both anthropogenic and climate change threats, establishing baseline data for these systems is urgent and essential. Data should include a range of variables relating to areal extent, degree of development, biodiversity, resilience, resource use and values, threats, dependence by local communities, critical species habitats, etc. with details specific to each site. These data provide the baseline from which to measure long-term trends and the ability of species, habitats, and ecosystems to cope with climate change.

If conservation strategies are to be successful in protecting species and habitats, they will need to adapt to the changing climate conditions. The ability to predict the location of future habitat sites, and build these potential sites into PA design and adaptation, will be a crucial element of long-term planning to ensure sustainable PAs in the face of global change. Flexible strategies and boundaries should be established and tracked to allow for adaptive management. Ecosystem management needs to be adaptive and informed by developing science, the findings of ongoing monitoring activities, and changes relative to established baselines. Monitoring programs measure change against indicators of ecosystem condition, uses, and levels of threat. Such programs provide useful time series data that can yield important information on ecosystem dynamics, long-term trends for key species, habitats, and resource uses, and their recovery response and potential following over-exploitation or severe damage from catastrophic events, including climate change impacts. As such, monitoring provides key information to inform adaptive management and can lead to changes in PA design and zoning and in management strategies, intensity, and geographic focus.

There is no substitute for direct field observations in tracking the course, impact and recovery from a perturbation. This will help PA managers to understand the vulnerabilities and diverse responses of key species and communities in different locations to climate change, to identify places of greater or lesser resilience, and review and revise PA zoning and management plans.

An on-going task is to develop sustainable livelihoods for the local people living in this harsh and remote environment. All too often, experience shows that biodiversity conservation can have detrimental effects on local peoples. Expanding the number and area of PAs increases the restrictions on people's activities, access, and resource uses. Effective management must therefore address the socioeconomic impacts of both the climate events themselves, and of the conservation measures introduced to counteract them. While local stewardship and sustainable harvest of alpine resources still offers great promise where strong traditional rights and management structures persist, traditional practices – including locally managed meadow areas – need to be nested within more robust and enduring governance frameworks to help ensure that they survive change. They also need to be buffered against increased demands for food and income by the development and introduction of alternative sustainable livelihoods as societies shift from subsistence based to cash economies.

Scientists and practitioners now recognize that governance and management frameworks should include diverse patterns of resource use to maintain social and ecological resilience (Adger et al. 2005). Alternative livelihood options and diverse income opportunities allow communities to be flexible to adapt to social, political, and economic changes.

## CONCLUSION

Given that climate will continue to stress ecosystems for decades to come, even were carbon emissions to be reduced from today onward, our only hope is to alleviate the impacts of climate change and implement actions that build resilience into these ecosystems. Strategically placed and well-managed PA networks that are specifically designed for resilience to climate change offer the most viable means of protecting and conserving key alpine species and ecosystems in perpetuity.

The development of PA networks that are resilient to the impacts of global climate change requires an integrated multidisciplinary approach that includes:

- collaboration among government agencies, conservation organizations, the private sector, and local communities
- policy reform to build flexibility into formal PA designations and require periodic review and revisions of their boundaries, zones and management strategies
- scientific research to increase understanding of resilience, including factors determining resistance and connectivity, and the principles guiding PA network design
- adaptive management of PAs to enable effective response to change in climate, demands, and pressures on the PAs
- increased levels of scientific support for locally managed meadow areas to ensure that they are located, designed, and managed to be resilient to climate change.

Brandt et al., 2009 reported that at least 39% of alpine meadows they studied had converted from meadows to shrubs in a study site in north-west Yunnan province between 1990 and 2009. A broader conversion occurring across the region from herbaceous to "shrub-dominated" ecosystems. Part of the reason, was that declining snow cover often gives shrubs a competitive advantage over herbaceous plants. Urgenson et al., (2014) charted a nearly 70% decline in meadow area from 1974 to 2004 in China's Jiuzhaigou national park, in Sichuan province. Climate change appeared to be one of several causes of meadow shrinkage, along with overgrazing by yaks.

Worldwide, changing climates and land use practices are escalating woody-plants encroachment into grasslands, reducing biodiversity and altering ecosystem functions. The loss of alpine grasslands is a major conservation concern as they harbor many rare and endemic species. Alpine meadows in Northwest Yunnan, China, represent a global biodiversity hotspot with high species richness, beta diversity, and endemism. Shrubs have expanded greatly in the region and threaten alpine meadow biodiversity. To measure rates of meadow loss due to shrub encroachment and identify its mechanisms, the researchers reconstructed alpine land cover, climate, and land use change from 1950 to 2009 across Northwest Yunnan using satellite data, ground surveys, and interviews. Between 1990 and 2009, at least 39% of the alpine meadows converted to woody shrubs. The patterns of change suggest that a regime shift is occurring. Despite multiple perturbations to the climate and land use systems starting in the 1950s, alpine meadows remained resilient to shrub expansion until the late 1980s. Shrublands rapidly expanded then due to feedback mechanisms involving climate, woody cover, and grazing. Fire may no longer be an effective tool for controlling shrub expansion. This regime shift threatens both endemic meadow biodiversity and local livelihoods. More generally, these trends serve as a warning sign for the greater Himalayan region where similar vegetation changes could greatly affect livelihoods, hydrology, and regional climate.

Shrub encroachment in grasslands, affects biodiversity, ecosystem function, and carbon storage (Knapp et al., 2008). Climate change, overgrazing, and land use all contribute to shrub encroachment (Walker and Wahren, 2006). When multiple interacting factors reach a critical threshold, a relatively small perturbation can trigger a regime shift, i.e., a sudden, and potentially irreversible, transition to an alternate state (Scheffer, 2009). To conserve grassland biodiversity, the drivers of shrub encroachment need to be identified and their critical thresholds determined. Untangling these factors, however, is challenging in coupled natural and human systems (Brock and Carpenter, 2010). Shrub encroachment into alpine meadows represents a particular grave conservation threat (Montane et al., 2007; Xu and Wilkes, 2004). Alpine meadows often harbor many endemic plants adapted to harsh conditions, and orogenies have resulted in high levels of both paleo- and neo-endemism (Wu et al., 2007).

With the largest alpine ecosystems in the world, alpine meadows in the Greater Himalayan region are of particular concern. Northwest west China represents a global biodiversity hotspot (Myers et al., 2000), and meadows here have higher species richness, beta diversity, and endemism than elsewhere in the Himalayas (Salick et al., 2009; Xu et al., 2009). Historically, Himalayan meadows have been used by indigenous agro-pastoralists whose rangeland management practices sustained both local livelihoods and biodiversity (Klein et al., 2011).

The highly nutritious meadows are critical to Tibetan transhumance herding, and the replacement of meadows by unpalatable shrubs threatens their livelihoods (Yi et al., 2007). Alpine meadow species can persist in the low-intensity land use systems typical for montane regions, but even subtle environmental changes may threaten them (Galop et al., 2011; Thuiller et al., 2005).

For example, when climate change shifts woody vegetation to higher elevations, species of alpine mountaintop meadows may not be able to migrate to other suitable environments (Parmesan, 2006). Likewise, the abandonment of traditional land use practices can result in shrub encroachment Laiolo et al., 2004). Indeed, climate change and land use change interact in alpine ecosystems making it difficult to identify the main driver(s) of shrub expansion.

## References and Further Reading


Adger, W.N.,T. P. Hughes, C. Folke, S. R. Carpenter, and J. Rockström 2005. Social-Ecological Resilience to Coastal Disasters. Science 309 (5737): 1036-1039 DOI: 10.1126/science.1112122

Brandt, J.S.,M.A. Haynes, T. Kuemmerle, D.M. Waller, and V.C. Radeloff 2013. Regime shift on the roof of the world: Alpine meadows converting to shrublands in the southern Himalayas. *Biological Conservation* 158 (2013) 116–127

Brock, W.A., Carpenter, S.R, 2010. Interacting regime shifts in ecosystems: implication for early warnings. *Ecol. Monogr.* 80, 353–367.

Diaz, HF, Grosjean M, Graumlich L. 2003. Climate variability and change in high elevation regions: past, present and future. *Climatic Change* 59: 1–4. doi: 10.1007/978-94-015-1252-7_

Dong, S. K., Gao, H. W., Xu, G. C., Hou, X. Y., Long, R. J., Kang, M. Y., and Lassoie, P. J. 2007. Farmer and professional attitudes to the large-scale ban on livestock grazing of grasslands in China. *Environmental Conservation*34, 246–254.

Dong, S. K., Li, J. P., and Li, X. Y. 2010. Application of design theory for restoring the 'black beach' degraded rangeland at the headwater areas of the Qinghai-Tibetan Plateau. *African Journal of Agricultural Research* 5, 3542–3552.

Duan, M. J., Gao, Q. Z., Wan, Y. F., Li, Y. E., Guo, Y. Q., Danjiu, L. B., and Luosang, J. C. 2010. Effect of grazing on community characteristics and species diversity of *Stipa purpurea* alpine grassland in Northern Tibet. *Acta Ecologica Sinica* 30, 3892–3900.

Fan, Y. G., Hu, Y. K., Li, K. H., Yu, J. M., and Wang, X. 2008. Effects of different disturbances on the diversity and biomass of the phytobiocoenoses in alpine steppes. *Arid Zone Research* 25, 531–536.

Galop, D., Houet, T., Mazier, F., Leroux, G., Rius, D., 2011. Grazing activities and biodiversity in the Pyrenees: new insight on high altitude ecosystems in the framework of a Human–Environment Observatory. *PAGES News* 19, 53–55.

Gao, Q. Z., Li, Y. E., Wan, Y. F., Lin, E. D., Xiong, W., Jiangcun, W. Z., Wang, B. S., and Li, W. F. 2006. Grassland degradation in Northern Tibet based on remote sensing data. *Journal of Geographical Sciences* 16, 165–173.

Gao, Q. Z., Li, Y., Wan, Y. F., Jiangcun, W. Z., Qin, X. B., and Wang, B. S. 2009. Significant achievements in protection and restoration of alpine grassland ecosystem in Northern Tibet, China. *Restoration Ecology* 17, 320–323.

Gao, Q. Z., Li, Y., Xu, H. M., Wan, Y. F., and Jiangcun, W. Z. 2014). Adaptation strategies of climate variability impacts on alpine grassland ecosystems in Tibetan Plateau. *Mitigation and Adaptation Strategies for Global Change* 19, 199–209.

Guisan, A, and Theurillat, J. P, 2000. Assessing alpine plant vulnerability to climate change: a modeling perspective. *Integrat Ass.* 1: 307–320.

Hampe, A, and R.J. Petit 2005. Conserving biodiversity under climate change: the rear edge matters. *Ecol Lett* 8: 461–467. doi: 10.1111/j.1461-0248.2005.00739.x

Harris, R. B. 2010. Rangeland degradation on the Qinghai-Tibetan plateau: a review of the evidence of its magnitude and causes. *Journal of Arid Environments* 74, 1–12.

Holling, C. S. 1973. Resilience and stability of ecological systems. *Annual Review of Ecology and Systematics* 4:1-23.

Houghton, J., Ding, Y., Griggs, D., Noguer, M., van der Linden, P., et al. (eds.). 2001. Climate Change 2001: The Scientific Basis. Published for the Intergovernmental Panel on Climate Change. Cambridge University Press, Cambridge, UK.

Klein, J., Yeh, E., Bump, J., Nyima, Y., and Hopping, K., 2011. Coordinating environmental protection and climate change adaptation policy in resource-dependent communities: a case study from the Tibetan Plateau. In: Ford, J.D. (Ed.), Climate Change Adaptation in Developed Nations. Springer.

Knapp, A.K., Briggs, J.M., Collins, S.L., Archer, S.R., Bret-Harte, M.S., Ewers, B.E., Peters, D.P., Young, D.R., Shaver, G.R., Pendall, E., and Cleary, M.B., 2008. Shrub encroachment in North American grasslands: shifts in growth form dominance rapidly alters control of ecosystem carbon inputs. *Glob. Change Biol.* 14, 615– 623.

Laiolo, P., F. Dondero, E. Ciliento and A Rolando 2004. Consequences of pastoral abandonment for the structure and diversity of the alpine avifauna. *Journal of Applied Ecology* 41: 294–304

Lenoir, J. and Gegout JC, Intergovernmental Panel on Climate Change. 2007. Summary for Policymakers. [18 pp.] in Climate Change 2007: The Physical Science Basis. Contribution of Working Group I to the Fourth Assessment Report of the Intergovernmental Panel on Climate Change [Solomon, S., Qin, D., Manning, M., Chen, Z., Marquis, M., et al. (eds.)]. Cambridge University Press, Cambridge, UK & New York.

Li, X.-L, G. L. W. Perry, G. Brierley, H.-Q. Sun, C.-H. Li and G.-X. Lu 2014. Quantitative assessment of degradation classifications for degraded alpine meadows (Heitutan), Sanjiangyuan, western China. *Land Degradation & Development* 417–427 DOI: 10.1002/ldr.2154

Liu X., and Zhang M. 1998. Contemporary climatic change over the Qing.hai-Xizang plateau and its response to the green-house effect. *Chinese Geogr Sci* 8: 289–298. doi: 10.1007/s11769-997-0034-9.

Liu, X. M., and Nie, X. M. 2012. Effects of enclosure on quantitative characteristics of alpine vegetation. *Acta Prataculturae Sinica* 29, 112–116.

Liu, W., Zhou, H. K., and Zhou, L. 2005. Biomass distribution pattern of degraded grassland in alpine meadow. *Grassland of China* 27, 9–15.

Ma, Y. S., Lang, B. N., Li, Q. Y., Shi, J. J. and Dong, Q. M. 2002. Study on rehabilitating and rebuilding technologies for degenerated alpine meadow in the Changjiang and Yellow River source region. *Pratacult. Sci.* (in Chinese). 19: 1–5.

Madsen, D.B., F. Chen,and X. Gao 2011. *Late Quaternary Climate Change and Human Adaptation in Arid China* Elsevier, -244 p.

Marquet, PA, de Ruffray P, and Brisse H 2008). A significant upward shift in plant species optimum elevation during the 20$^{th}$ century. *Science 320*: 1768–1771. doi: 10.1126/science.1156831

Miehe, G 1996) On the connexion of vegetation dynamics with climatic changes in high Asia. Palaeogeogr Palaeocl 120: 5–24. doi: 10.1016/0031-0182(95)00031-3.

Montane, F., Rovira, P., and Casals, P., 2007. Shrub encroachment into mesic mountain grasslands in the Iberian peninsula: effects of plant quality and temperature on soil C and N stocks. *Global Biogeochem. Cycl.* 21.

Myers, N., Mittermeier, R., Mittermeier, C., da Fonseca, G.A.B., and Kent, J., 2000. Biodiversity hotspots for conservation priorities. *Nature* 403, 853–858.

Parmesan, C. 2006. Ecological and evolutionary responses to recent climate change. Ann Rev Ecol Evol Syst 37: 637–669. doi:

Salick, J, F. Zhendong, and A. Byg 2009. Eastern Himalayan alpine plant ecology, Tibetan ethnobotany, and climate change. *Global Environ Change* 19: 147–155. doi: 10.1016/j.gloenvcha.2009.01.008

Scheffer, M., 2009. *Critical Transitions in Nature and Society*. Princeton University Press, Princeton, NJ.

Shang, Z. H., Gibb, M. J., Leiber, F., Ismail, M., Ding, L. M., Guo, X. S., and Long, R. J. 2014). The sustainable development of grassland-livestock systems on the Tibetan plateau: problems, strategies and prospects. *The Rangeland Journal* 36, 267–296.

Shang, Z. H., Ma, Y. S., Long, R. J. and Ding, L. M. 2008. Effect of fencing, artificial seeding and abandonment on vegetation composition and dynamics of 'black soil land' in the headwaters of the Yangtze and the Yellow Rivers of the Qinghai-Tibetan Plateau. *Land Degrad. Develop.* 19: 554–563.

Shi, F. S., Wu, N., Luo, P., Yi, S. L., Wu, Y., Wang, Q., and Gao, Y. H. 2007. Effect of enclosing on community structure of subalpine meadow in Northwestern Sichuan, China. *Chinese Journal of Applied and Environmental Biology* 13, 767–770.

Shrestha UB, Gautam S, Bawa KS 2012) Widespread climate change in the Himalayas and associated changes in local ecosystems. PLoS ONE 7(5): e36741 doi:10.1371/ journal.pone.0036741.

Squires, V.R. (in press) Sustainable rangeland management: an ecological and economic imperative. pp.3-18 In: Victor R Squires (Ed) Rangeland Ecology, Management and Conservation Benefits, NOVA Press, N.Y.

Squires, V., X. Lu, Q. Lu, T. Wang, Y. Yang 2009. Rangeland Degradation and Recovery in China's Pastoral Lands. CABI, Wallingford 264 p.

Squires, V.R. and N.M. Safarov, 2013. Diversity of plants and animals in mountain ecosystems in Tajikistan. *J. Rangeland Sci.* 4: 43-60

Sun, L., Wang, X. H., and Wei, X. H. 2012. Effect of closed fencing on the restoration of deteriorated alpine grasslands in Northern Tibet Plateau. *Journal of Anhui Agricultural Sciences* 40, 12072–12074.

Telwala, Y, Brook BW, Manish K, Pandit MK 2013. Climate-Induced Elevational Range Shifts and Increase in Plant Species Richness in a Himalayan Biodiversity Epicentre. *PLoS ONE* 8(2): e57103. doi:10.1371/journal.pone.005710

Thuiller, W., S. Lavorel, M. Araujo, M.T. Sykes, and I.C. Prentice. 2005. Climate change threats to plant diversity in Europe. *Proc. Natl. Acad. Sci.* USA 102:8245–50.

Urgenson, L., A. H. Schmidt, J. Combs, S. Harrell, T. Hinckley, QX Yang, Z. Ma, Y. Li, H. Lü, and A. MacIver. 2014. Traditional Livelihoods, Conservation and Meadow Ecology in Jiuzhaigou National Park, Sichuan, China. *Human Ecology* 42(3):481-491.

Walker, W., Harremoës, P., Rotmans, J., van der Sluijs, J., van Asselt, M., Janssen, P. & Krayer von Krauss, M. 2003. 'Defining uncertainty: A conceptual basis for uncertainty management in model-based decision support', *Integrated Assessment* 4(1):5–18.

Walker, B., Carpenter, S. Anderies, J. Abel, N. Cumming, G. S. Janssen, M. Lebel, L. Norberg, J. Peterson, G. D. and Pritchard, R. 2002. Resilience management in social-ecological systems: a working hypothesis for a participatory approach. *Conservation Ecology* 6(1): 14.

Walker, M., and Wahren, C., 2006. Plant community responses to experimental warming across the tundra biome. *Proc. Nat. Acad. Sci. U.S.A.* 103, 1342–1346.

Walther, GR, Beißner S, Burga CA 2005) Trends in the upward shift of alpine plants. *J. Veg. Sci.* 16: 541–548. doi: 10.1658/1100-9233(2005)16[541:tituso]2.0.co;2.

Wu, G. L. and Du, G. Z. 2007. Discussion on ecological construction and sustainable development of degraded alpine grassland ecosystem of the Qinghai Tibetan Plateau. *Chinese J. Nat.* (in Chinese). 29: 159–164.

Wu, G. L., Du, G. Z., Liu, Z. H. and Thirgood, S. 2009. Effect of fencing and grazing on a Kobresia-dominated meadow in the Qinghai-Tibetan Plateau. *Plant & Soil.* 319: 115–126.

Wu, G. L., Li, W., Zhao, L. P. and Shi, Z. H. 2011. Artificial management improves soil moisture, C, N and P in an alpine sandy meadow of western China. *Pedosphere.* 21(3): 407–412.

Wu, G. L., Liu, Z. H., Zhang, L., Cheng, J. M. and Hu, T. M. 2010a. Long-term fencing improved soil properties and soil organic carbon storage in an alpine swamp meadow of western China. *Plant & Soil.* 332: 331–337.

Wu, G. L., Liu, Z. H., Zhang, L., Hu, T. M. and Cheng, J. M. 2010b. Effects of artificial grassland establishment on soil nutrients and carbon properties in a black-soil type degraded grassland. *Plant &Soil.* 333: 469–479.

Wu, G. L., Ren, G. H., Liu, Z. H., Yang, L. P. and Banma, C. R. 2009b. Restoration and improvement of "black soul beach" degraded meadows in Maqu wetland function area of Yellow River. *J. Lanzhou Univ. Nat. Sci.* (in Chinese). 45: 48–52.

Wu, J. S., Zhang, X. Z., Shen, Z. X., Shi, P. L., Yu, C. Q., Song, M. H., and Li, X. J. 2012. Species richness and diversity of alpine grasslands on the Northern Tibetan Plateau:

effects of grazing exclusion and growing season precipitation. *Journal of Resources and Ecology* 3, 236–242.

Wu, Q. H., Li, J. Q., Zhang, F. W., Liu, X. Q., Mao, S. J., and Li, Y. N. 2013. Distribution patterns of vegetation-and soil carbon and nitrogen density in an alpine forb meadow under short-term grazing gradient. *Chinese Journal of Ecology* 32, 2857–2864.

Xiong, DP., P. Shi, Y. Sun, J. Wu, X. Zhang 2014. Effects of grazing exclusion on plant productivity and soil carbon,, nitrogen storage in alpine meadows in northern Tibet, China. *Chinese Geographical Science* 24 (4): 488-498.

Xu, J., and Wilkes, A., 2004. Biodiversity impact analysis in northwest Yunnan, southwest China. *Biodivers. Conserv.* 13, 959–983.

Xu, L. L., Zhang, X. Z., Shi, P. L., Yu, G. R., and Sun, X. M. 2005. Net ecosystem carbon dioxide exchange of alpine meadow in the Tibetan Plateau from August to October. *Acta Ecologica Sinica* 25, 1948–1952.

Xu, J., R.E.Grumbine, A. Shrestha,, M. Eriksson, X. Yang, Y. Wang, and A. Wilkes, 2009. The melting Himalayas: cascading effects of climate change on water, biodiversity, and livelihoods. *Conserv. Biol.* 23, 520–530.

Yi, S., Z. Zhou, S. Ren, M. Xu, Y. Qin, S. Chen and B. Ye 2005. Effects of permafrost degradation on alpine grassland in a semi-arid basin on the Qinghai–Tibetan Plateau. *Environ. Res. Lett.* 6 (2011) 045403 (7pp) doi:10.1088/1748-9326/6/4/045403

Zhang, W.N., H. Ganjurjav, Y. Liang, Q. Z. Gao, Y. F. Wan, Y. Li, Y. Z. Baima and Z. M. Xirao 2015. Effect of a grazing ban on restoring the degraded alpine meadows of Northern Tibet, China *The Rangeland Journal* 37(1) 89-95 http://dx.doi.org/10.1071 /RJ14092

Zhang, Y., Wu, G. L., Ren, G. H. and Liu, Z. H. 2009. Effects of reseeding 'alpine No # 1' eco-grass on degraded alpine meadows in Maqu after enclosure. *Pratacult. Sci.* (in Chinese). 26: 9.

Zhou, H. K., Zhao, X. Q., Zhou, L., Liu, W., Li, Y. N., and Tang, Y. H. 2004. A study on correlations between vegetation degradation and soil degradation in the 'Alpine Meadow'of the Qinghai-Tibetan Plateau. *Acta Prataculturae Sinica* 14, 31–40.

In: Rangeland Ecology, Management ...
Editor: Victor R. Squires ISBN: 978-1-63482-504-7


*Chapter 10*

# UNIFYING PERSPECTIVES: ENSURING CONSERVATION BENEFITS THROUGH BETTER RANGELAND MANAGEMENT

***Victor R. Squires***
Gansu Agricultural University, Lanzhou, China

## ABSTRACT

Rangelands, on a global basis, are thought of principally as constituting a pastoral resource and rangeland science aims at a better understanding of how that resource may be more sustainably and efficiently utilized by land users, pastoralists and their animals, and by conservationists who wish to preserve biodiversity and eco-tourists who wish to enjoy the wilderness experience. Issues in rangeland are usually complex, generally operate over multiple scales and are continuously changing. Clearly multiple-use is a desired goal. In spite of a strong emphasis on pastoral use, there is growing recognition that rangelands are used by people in many ways in addition to production of animal products by pastoralism. Rangeland management must cope with highly unpredictable and variable rainfall extremes.

**Keywords**: conflict, landscape function, socio-economic, ecology, biophysical, climate change, China, biodiversity, poverty, multiple-use

## 1. INTRODUCTION

In this final chapter there is an attempt to summarize the main threads of thinking and issues for rangeland ecology and management in the context of the conservation benefits society will achieve through better stewardship of these important resources. A number of important themes are implicit in the title of the book '*Ecology*, '*Management*' and '*Conservatio*n'. Each of these is important in its own right. An understanding of the ecology of rangelands underpins the management and provides a direction for efforts to conserve the rangelands for the benefit of current and future generations. Much has been written on these subjects since range science became a discipline of study in its own right in the 1920-30s (Grice and Hodgkinson, 2002). Rangeland ecosystems can be defined as the places where people live and in which they have an integral role. Not all the people are involved in pastoral activities. Clearly multiple-use is a desired goal. In spite of a strong emphasis on pastoral use, there is growing recognition that rangelands are used by people in many ways in addition to production of animal products by pastoralism. A number of examples can be given. Mining industries are prominent in many parts of the world's rangelands. Although utilizing only a very small part of the world's rangelands, the mining industries are economically very important, directly and indirectly, for the sustainability of rangeland people and their businesses. Mining does not specifically require a rangeland environment, but where it occurs there may be substantial direct and indirect economic benefits to rangeland people. Mining can also be environmentally destructive (Figure 1) and rehabilitation of mined areas is a challenge.

Figure 1. Uncontrolled mining on rangelands, like these examples from Mongolia can be very destructive.

In Mongolia for example, mining licence-holding companies turned soil of over 10,000 hectares of land into tailings, conducted mechanical reclamation on less than 30 per cent of eroded land, however, no biological reclamation is done. Alongside with mining companies, there are 15- thousand illegal "ninja" miners (Figure 1) who mine gold, fluorspar and other metals by hand and it is impossible to estimate the damage they cause by digging land and degrading its productivity.

The inspection carried out by Environment Ministry in 2009 at 34 mines in Mongolia's western region, 36 mines in central region, 37 mines in eastern region, total of 107 companies and organizations engaged in mining, showed that in these territories, 28,510.9 $10^3$ cubic meters of topsoil was removed and 25,032.52 2 $10^3$cubic meters of sand was mined. Gold mines used in 2004 76.8 $10^6$ cubic meters of water, in 2005 96.8 million cubic meters, in 2006 93.8 $10^6$cubic meters. Most mines don't follow mining standards, do not do biological reclamation. Out of a total land area utilized by these 107 mining companies, no reclamation is done at all on 55 per cent or 2454 hectares. The tailings of gold mining companies degrade significantly productivity of soil and cause double damage to environment, they also close (without any permission) flows of rivers, change their regime, swamp land with dirty water, dig holes and gorges and cause great danger to the natural environment.

Biological and ecosystem assets are the foundation of the survival of the rural poor (see below for a discussion of link between poverty and biodiversity conservation) Biodiversity conservation is an essential component of poverty elimination within the rural context. Rangelands, on a global basis, are thought of principally constituting a pastoral resource and rangeland science aims at a better understanding of how that resource may be more sustainably and efficiently utilized by land users, pastoralists and their animals, and by conservationists who wish to preserve biodiversity and eco-tourists who wish to enjoy the wilderness experience (Figure 2).

A much deeper understanding has led to changes in people's thinking (Behnke, Scoones and Kerven, 1993; Ludwig et al., 1997; Heshmati and Squires, 2009). There is now a better understanding of the principal objectives of rangeland management (Figure 3).

The landscape function approach requires a major shift in thinking about rangeland management. If does not ask about, for example, how do we shape rangelands to maximize output of a commodity like meat, wool or mohair? Instead it turns the question around and asks how a semi-arid landscape functions under consumption and trampling pressure. It asks for grazing management practice which improve the conservation of scarce water and nutrients, and hence yield of higher forage and livestock production, offtake and income. The integration of management options to achieve 'ecologically sustainable development' is a major challenge facing rangeland users and managers. This task is made considerably more daunting in arid and semi-arid rangelands given their inherently unpredictable climate and spatial variability.

Rangeland management must cope with highly unpredictable and variable rainfall extremes. During climatic extremes opportunities for management intervention exist. For example, droughts provide opportunity to deal with feral animals like camels (Zeng, this volume) wild pigs, feral donkeys etc. or change livestock type. Surplus forage produced after 'big wets' can be used as fuel for controlled burning in the fire management program (Peter et al., this volume).

Figure 2. The vast rangelands in arid regions provide ecotourism opportunities as the unexpected can show up around the next bend in the road.

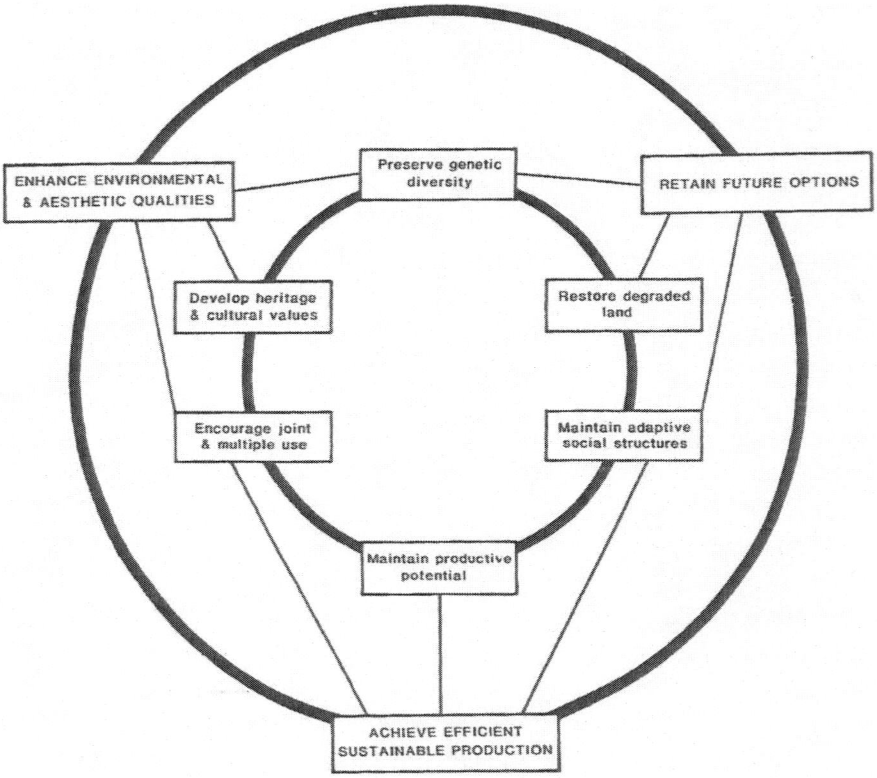


Figure 3. Management aims and objectives for the rangelands.

Rangeland managers, land users and policy makers are facing ever increasing conflicts over the demands and values being placed on using these lands for different purposes. Resolution of these conflicts is easier if one understands how such systems operate. If conflicts over the different uses of rangelands are to be resolved it is imperative that we understand how they function, and how they work (Ludwig et al., 1997, Tongway and Ludwig, 2011).

In any debate, aimed at resolving conflicts concerning multiple-use of rangelands (pastoralism, mining, conservation and biodiversity reserve, military training areas, eco-tourism, watersheds etc.) there is need for a better understanding of the basic ecology of such lands. An ecological approach can help with this understanding and begin by developing a conceptual framework for describing how arid and semi-arid landscapes function.

By definition, when rangeland landscapes become increasingly dysfunctional their capacity to capture and store scarce soil-water and nutrients declines. They tend to lose accumulated reserves. Instead there is an accelerated loss of these resources out of the landscape to other systems (e.g., loss of soil and nutrients to rivers and lakes. There are no 'traps' to catch and retain water, sediments and plant litter (organic carbon).

Landscape function analysis, as well as indicating current functional status, also provides clues as to whether a particular level of dysfunction can be reversed by simply removing the cause of the degradation (Tongway and Ludwig, 2011). The net development of a rehabilitated rangeland toward becoming a highly functional, self-sustaining, dynamically balanced system initially depends on applying technologies to improve physical processes (e.g., water retention) but, over time, the role of these processes is largely taken over by self-replacing biological processes (e.g., plant growth and reproduction).

## 2. Achieving Ecologically Sustainable Development (ESD) in Rangelands

ESD will be difficult because land users expect to make a living off the land. At the bigger scale there is clear need to integrate ecological and social sciences and develop a classification of rangelands into appropriate regions in order to address inter-regional questions of production potential, conservation value, resilience and socio-economic viability. Rangelands science has commonly only dealt with biophysical constraints. There have been numerous reminders that factors, other than biophysical ones also limit management options and the sustainability of rangeland enterprises, industries and communities. It is now recognized that socio-economic factors are often crucial in determining whether particular technical factors can be employed in employed in practice.

At a regional scale, processes and tools need to be developed to optimize the multiple uses of rangelands (i.e., finding the preferred mix). Procedures need to be developed by scientists and stakeholders to integrate ecological and societal needs and *inter alia* achieve conservation benefits. A common problem arises when the need of pastoralists to remain economically viable conflicts with the desires of society and scientists to conserve biodiversity within a region. One approach is to classify the different major rangeland types within the region by the conservation value and economic value (Figure 3). No problems are likely to occur over those areas with high conservation value and low economic viability.

Areas with low economic viability and low conservation value may cause some problems but these are likely to be minor. However, trade-off and conflict resolution procedures will be needed integrate land uses in those areas with both high conservation value and high economic viability.

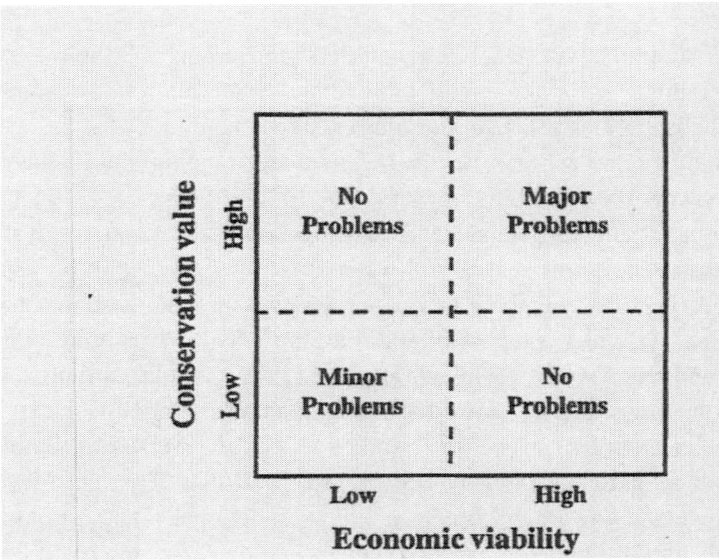


Figure 3. A compartmentalization of the problems likely to occur when the goal is to achieve economic viability from rangelands, whilst also achieving conservation goals.

In our changing world, human populations are rapidly growing and demanding more of the goods and services (Squires, this volume) to be provided by the rangelands.

Many rangelands are degrading (Squires et al., 2009) and the level of functional improvement aimed for depends on the goals of the stakeholder – those people with an interest in, or dependence on, a rangeland landscape, whether or not they live in the rangeland. Stakeholders may wish to restore a rangeland to a more natural state to achieve improved biodiversity goals (Escribano et al., this volume).

## 3. Poverty Alleviation, Biodiversity and Environmental Stewardship

Contributions are made both to conservation and to poverty reduction when we support the poor in using these assets sustainably so that they are maintained or increased. Ecosystems are assessed in terms of well-being and quality of human life by their ability to provide services in the form of available resources, regulation of components and processes and intangible services (e.g., esthetic enjoyment).

Resource degradation affects the rural poor when the quality of the resources they depend upon are seriously affected so that they no longer perform their ecosystem function or support human welfare. The rural poor are highly vulnerable to many of the same actions that threaten biodiversity. In particular, the poor are most reliant on natural resources for food, shelter, medicine, income and employment. If the rural poor draw their livelihoods from a fragile and steadily degrading ecological base, there is no potential for sustainable lives and livelihoods. At the same time, it is vital to also acknowledge that in most fragile ecosystems, the poor are unlikely to escape from poverty by sustainably exploiting biodiversity. Others ancillary measures must be taken.

Actions to conserve biodiversity also contribute to maintaining intact habitats and ecological processes that support human welfare by protecting the core ecosystem services (Table 2 p. 6 in Squires, this volume) that are the foundation for sustainable and healthy human lives.

## 4. Measuring Change in Rangelands

New tools are needed to effectively apply sustainable principles to planning and management. The spatial dimension of sustainability engages processes and relations between different land uses, ecosystems and biotopes at different scales, and over time. Although the economical dimension of sustainability is arguably assured its landscape, ecological and territorial dimensions are not entirely.

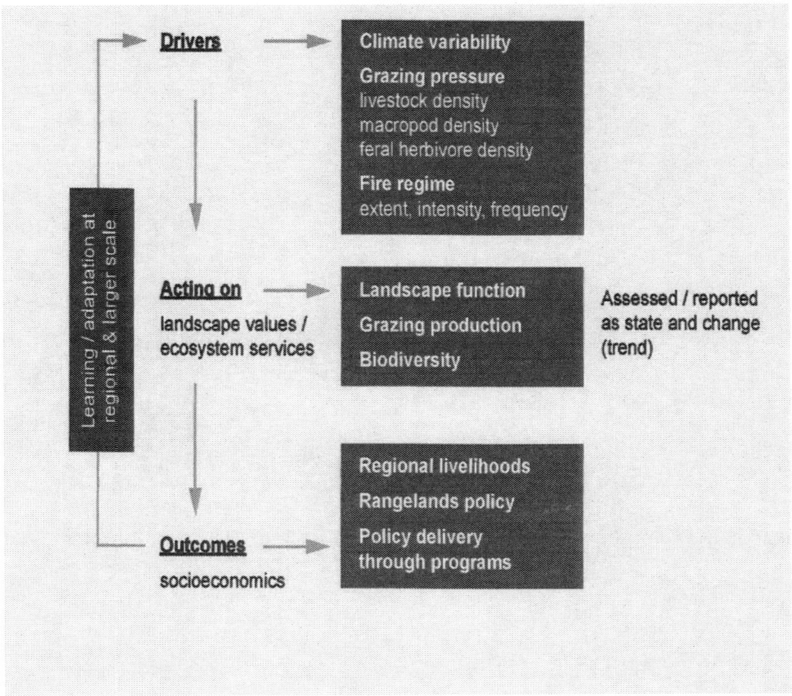


Figure 4. Flow chart of linkages between drivers of change that act on landscape values and ecosystem services.

Tools associated with reporting on the impact of key ecosystem drivers have been developed. It is structured around a simple framework that relates key drivers of change to effects on landscape values or ecosystem services. These effects, in turn flow through to socio-outcomes. These outcomes feedback on the drivers, ideally though appropriate learning and adaptive management.

Important questions are: What's happening in complex environments from local to global level? What's important to different land users and stakeholders? Indicators have been developed and the principal attributes of successful indicators are shown on Box 1.

**Box 1. Characteristics of a good environmental indicator**

- Has an agreed, scientifically sound meaning
- Represents an environmental aspect of importance to society
- Tells us something important, and its meaning is readily understood
- Has a sound and practical measurement process
- Helps focus information to answer important questions
- Assist decision making by being effective and cost-efficient to use

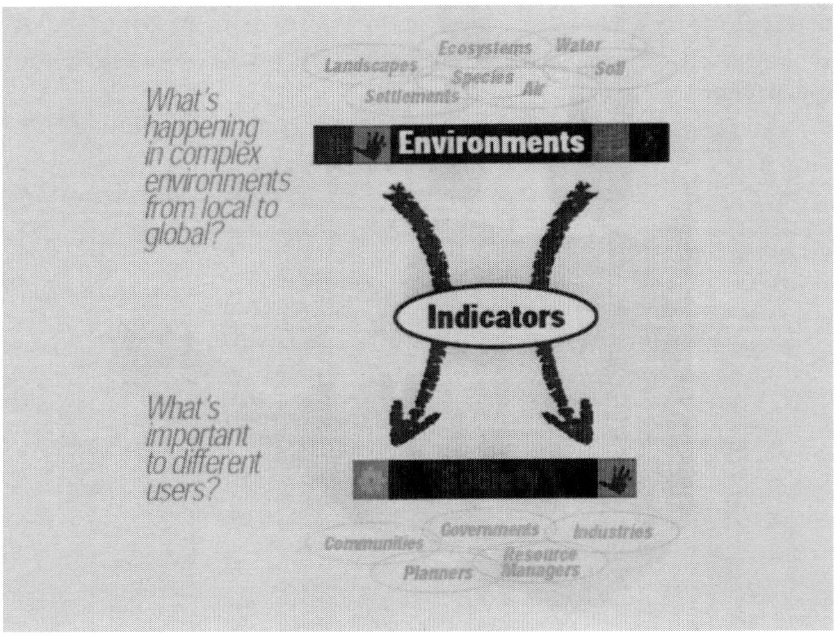


Figure 6. Indicators are important tools to assess what is happening in the environment and among the land users.

## 5. Institutional Capacity for Change

In both developed and developing countries, the institutions which are charged with, or assume, management and development responsibilities with respect to rangeland ecosystem and societies have often done badly in the past 70-100 years and are often poorly equipped to deal with in the decades to come. For instance, the degradation crisis identified repeatedly over the past 100 years remain unresolved in spite of dozens of inquiries and he development of a plethora of institutional structure can even generate an 'institutional gridlock' crisis (where competing government agencies and their jurisdictional statutes often make wasteful and contradictory policy. In China (Hua et al., 2015) successive institutional change from nomadic to state control to privatization of livestock and private control over capital investment) has tended to intensify rangeland use but without recognition of the environmental and socio-economic constraints under which rangelands operate, resulting in marginalization of pastoral communities, increasing inequities and decreasing resource condition (Hua and Squires, 2015). The new sedentary pastoral system has also lost the flexible mechanism that enables effective response to environmental changes. Furthermore, vulnerability to change may be increased by replacing a diverse multi-resource economy with a single–product ranching system with rigid marketing and prices that do not reflect the true cost of production. In other regions, particularly in the nations which separated from the former USSR, there has been a demise of institutions which previously imposed strict controls on land use and management but also provided support through fodder production, financial backing and subsidized inputs (Squires, 2012). The evolution of new intuitional and other arrangements provides an opportunity for these rangelands regions to be more attuned to the needs of a globalized world: however, in the process, there have been considerabe human costs, increased inequities and resource degradation. The capacity to adjust to future change is uncertain given low levels of capital, degrading infrastructure such as roads and wells, the lack of scientific and technical groups and the uncertain tenure arrangements. In all of these cases, authors have expressed the need for reform to develop locally based institutions which use and enhance local indigenous knowledge (LEK) that recognize cultural, socio-economic and environmental factors that transfer rights and responsibilities to the local institutions and that develop the meeting places, the language and the processes whereby regional proposals, issues and disputes might be explored and resolved (Hua and Squires, this volume). However, institutions are generally based on precedent not foresight and in a rapidly changing world there is the likelihood that they will become outdated as markets, cultures, and environments change (Squires 2012b, UNCCD, 2005). Thus an adaptive process needs to be incorporated which can allow for such change. Classical, theoretical reductionist approaches (single factor reasoning) based on ecological factors alone are likely to make a limited contribution to such processes. Instead what is required is a more inclusive, comprehensive framework for analysing rangeland issues which incorporates social, economic and ecological aspects and which provides pathways for implementation of the decisions made by local institutions (Hua and Zhang, 2012).

## CONCLUSION

New visions and prospects are emerging. Communities and individuals are increasingly developing their own visions in response to local and regional issues. Issues are usually complex, generally operate over multiple scales and are continuously changing. The visions therefore need to be adaptive so that lessons can be learnt from successes and failures and from changing internal and external environments (Squires, 2012 c). Issues cannot be universal; each region or area has its own social people, and resource characteristics. At the moment people and governments are experiencing difficulty in thinking about complex issues in a holistic manner. However, there are examples around the world of communities developing their own plans for action. They are seeking to ensure the conservation benefits that better rangeland management can bring. Rangeland people will shape their own futures; it is likely that science will be only a small part of this process.

The obvious task for rangeland professional is to produce management tools in collaboration with rangeland users. These tools will enable their communities to develop more inclusive, comprehensive frameworks for analysing rangeland management issues and planning appropriate responses. The frameworks must encompass the social, economic and biophysical systems and identify the pathways for implementation of the decisions made by the local institutions and groups. If rangeland science can address these issues it will contribute materially to diminishing the effects of marginalization, developing adequate life-styles and living infrastructures, reducing poverty and improving education while maintaining the traditional role of developing resilient systems which maintain resource function.

## REFERENCES

Behnke, R.H. Jr, I, Scoones and C. Kerven (Eds). 1993 *Range ecology at disequilibrium: New models of natural variability and pastoral adaptation in African savannas*. ODI Publications, Regent's College, Inner Circle, Regent's Park, London 248 pages

Escribano, A.J, P. Gaspar, F.J. Mesías, M. Escribano and F. Pulido (in press) Comparative Sustainability Assessment of Extensive Beef Cattle Farms in a High Nature Value Agroforestry System. pp.65-86, this volume

Grice, A.C. and K.C. Hodgkinson (Eds) 2002. *Global rangelands: Progress and Prospects*. CABI publishing in association with CIFOR. 299 p.

Heshmati, G.A. and V.R. Squires 2009 New thinking in range ecology, In: *Range and Animal Sciences and Resources Management*, [Ed. Victor R. Squires], *Encyclopedia of Life Support Systems (EOLSS),* Developed under the Auspices of the UNESCO, EOLSS Publishers, Oxford, UK

Hua, L.M. and V.R. Squires, 2015 Managing China's pastoral lands: current problems and future prospects. *Land Use Policy* 42: 129-137.

Hua, L.M. and Zhang, D. 2012. Engaging with land users: The first steps on a long road. pp. 333-356 In: V.R. Squires, (ed) *Rangeland Stewardship in Central Asia: Balancing Improved livelihoods, Biodiversity Conservation and Land Protection*. Springer, Dordrecht,

Hua, L.M., S.W. Yang, V.R. Squires, and G.Z. Wang. 2015 An Alternative Rangeland Management Approach in an Agro-Pastoral Area in Western China. *Rangeland Ecology and Management.* 68(2):109-118 dx.DOI.org/10.1016/j.rama 2015.01.001

Ludwig, J., D. Tongway, D. Freudenberger, J. Noble and K. Hodgkinson (Eds.) 1997. *Landscape Ecology Function and Management: Principles from Australian Rangelands.* CSIRO, Melb. 158 p.

Peter, G. F. A. Funk, M.A. Luna, S.S. Torres Robles (in press) Changes in Vegetation and in the Organic Soil Fractions Related with Grazing and Fire Histories in the Patagonian Monte, Argentina. pp. 87-110, this volume

Squires. V., X. Lu, Q. Lu, T. Wang and Y. Yang (Eds) 2009. *Rangeland Degradation and Recovery in China's Pastoral Lands*, CABI, Wallingford, UK

Squires, V.R. (Ed) 2012a *Rangeland Stewardship in Central Asia: Balancing Livelihoods, Biodiversity conservation and Land protection.* Springer, Dordrecht

Squires V.R. 2012b Governance and the Role of Institutions in the Central Asian Region. pp. 275-303 In: Squires, V.R. (ed) *Rangeland Stewardship in Central Asia: Balancing Improved livelihoods, Biodiversity Conservation and Land Protection.* Springer, Dordrecht

Tongway, D.J, and J.A. Ludwig, 2011. *Restoring Disturbed Landscapes: Putting Principles into Practice.* Island Press/Society for Ecological Restoration, Washington 189 p.

UNCCD, 2005. Promotion of Traditional Knowledge: A compilation of UNCCD Documents and Reports from 1997-2003. Bonn, Germany 159 pp.

Zeng, BX (in press) Preliminary Investigation of the Decomposition of Feral Camel Carcasses in Central Australia pp. 149-161, this volume.

# About the Editor

**Dr. Victor Squires** is an Australian who as a young man studied animal husbandry and rangeland ecology. He has a PhD in Rangeland Science from Utah State University, USA. He is former Dean of the Faculty of Natural Resource Management at the University of Adelaide, Australia where he worked for 15 years after a 22-year career in Australia's leading research organization (CSIRO). He is author/editor of 11 books including the 'Livestock Management in the Arid Zone' 'Range and Animal Sciences and Resources Management', 'Rangeland Stewardship in Central Asia: Balancing Improved Livelihoods, Biodiversity Conservation and Land Protection' numerous research papers on aspects of range/livestock relations. Since retirement from the University of Adelaide Dr. Squires was a Visiting Fellow in the East West Center, Hawaii, and is currently an Adjunct Professor in the University of Arizona, Tucson USA and at the Gansu Agricultural University, Lanzhou, China. He has been a consultant to World Bank, Asian Development Bank and various UN agencies in Africa, China, Central Asia and the Middle East. He was awarded the 2008 International Award and Gold Medal for International Science and Technology Cooperation by the Government of China and in 2011 was awarded the Friendship Award by the government of China. The Gold Medal is the highest award for foreigners. In 2015, Dr. Squires was honored by the Society for Range Management (USA) with an Outstanding Achievement Award.

# List of Contributors

**Banzaragch Tsesed**, Ministry of Environment and Green Development, Ulaanbaatar, Mongolia

**Bolormaa, Damindsuren**, Biologist, Research Institute of Animal Husbandry, Ulaanbaatar, Mongolia

**Chantsallkham Jamsranjav**, Mongolian Society for Range Management, Ulaanbaatar, Mongolia

**Escribano, A.J.** Department of Animal Production and Food Science, University of Extremadura, Spain

**Escribano, M. F**. Department of Animal Production and Food Science, University of Extremadura, Spain

**Funk, Flavia Alejandra**, Sede Atlántica, Universidad Nacional de Río Negro, Viedma, Argentina

**Gaspar, P**. Department of Animal Production and Food Science, University of Extremadura, Spain

**Hua Limin,** Rangeland Ecology and Management, Gansu Agricultural University, Lanzhou, China

**Kurbanova, Barno** Giprozem 15, Dushanbe, Tajikistan

**Luna,** Martín Alejandro, Sede Atlántica, Universidad Nacional de Río Negro, Viedma, Argentina

**Madaminov, A.A**. Institute of Botany, Dushanbe, Tajikistan

**Mesías, F.J.** Department of Animal Production and Food Science, University of Extremadura, Spain

**Peter, Guadalupe,** Sede Atlántica, Universidad Nacional de Río Negro, Viedma, Argentina

**Pulido,F.** Department of Animal Production and Food Science, University of Extremadura, Spain

**Squires, V.R**. Visiting Professor in Rangeland Ecology and Management, Gansu Agricultural University, Lanzhou. China

**Sukhtulga, Ts** National University of Mongolia, Mongolia

**Terendash S.** Research Institute of Animal Husbandry, Ulaanbaatar, Mongolia

**Torres Robles, Silvia Susana**, Sede Atlántica, Universidad Nacional de Río Negro, Viedma, Argentina

**Zeng, Benxiang**, Northern Institute, Charles Darwin University, Alice Springs, NT, Australia

**Zhang Degang,** Rangeland Ecology and Management, Gansu Agricultural University, Lanzhou, China

# INDEX

## A



## B



## C

## K



## L



## M



## N



## O



## P



## Q

## R



## S



## T



## U

## V



## W



## X



## Y



## Z